AUTOMATIC TRANSMISSIONS

AUTOMATIC
TRANSMISSIONS

WALTER B. LAREW

Chilton Book Company
Radnor, Pennsylvania

DEDICATED TO MY PARENTS AND FAMILY
without whose help and understanding
this book would not have been written

$\mathbb{P}$reface

AUTOMATIC TRANSMISSIONS usually are given trade names such as Hydra-Matic, TorqueFlite, or Cruise-O-Matic. Many transmissions bearing the same name differ from one another. The primary difference may be in the sizes of the components, which are those that are suitable for use with engines of various power and torque capacities. Sometimes the differences are greater than these, involving basic types of clutches in the gear system (such as fluid and mechanical clutches), numbers of elements in the fluid torque converters, or other design features.

The purpose of the author in this book is to discuss the principles underlying automatic transmissions. Those employed in existing transmissions provide a basis for the discussion. Certain types of transmissions were selected for study which differ from one another in basic ways. However, because of the factors above, the illustrations which pertain to power flow, the numbers of gear teeth used in the computations, and other such items should not be considered as being applicable to any particular model of the indicated type of transmission. In fact, they may not be applicable to any of the models. In many instances the numbers of gear teeth used in the computations are based on the gear ratios of a particular model. Likewise, the torque factors of the torque converters are illustrative, as are the speeds at which gear shifts occur and other such items.

W. B. LAREW

Contents

Preface   vi

List of Illustrations   xv

List of Tables   xvii

List of Charts   xviii

1. Introduction
1. Principles of Simple Gear Systems   1
2. Speed and Torque Factors   7
3. Work, Power, Energy, Etc.   9

2. Automatic Transmissions
1. Functions of Transmissions   16
2. Manually Controlled Transmissions   20
3. Automatic Transmission Components   21
4. Torque and Speed Characteristics   21
5. Fluid Clutches   24
6. Fluid Torque Converters   39

3. Planetary Gear Sets and Systems
1. Planetary Gear Sets   53
2. Planetary Gear Systems   54

4. Simple Planetary Gear Systems Employing Simple Gear Set With Ring Gear
1. Numbers of Gear Teeth   55
2. Conditions of Operation   57
3. Condition 1   57
4. Condition 2   58
5. Condition 3   58
6. Condition 4   59

7. Condition 5   59
8. Condition 6   60
9. Condition 7   61
10. Conditions 8–12   61
11. Condition 13   61
12. Conditions 14–18   62
13. Summary and Graphs   62

5. Compound Planetary Gear Systems Employing Simple Gear Set With Ring Gear

1. Conditions of Operation   66
2. Torque Relationships   67
3. Condition 1   69
4. Condition 2   72
5. Conditions 3 and 5   74
6. Conditions 4 and 6   75
7. Values of A and U   76
8. Summary and Graphs   80

6. Simple Planetary Gear Systems Emloying Compound Gear Set With Ring Gear

1. Conditions of Operation   83
2. Conditions 1–24   83
3. Conditions 25–36   87
4. Conditions 37–48   88
5. Conditions 49–60   88
6. Summary and Graphs   88

7. Compound Planetary Gear Systems Employing Compound Gear Set With Ring Gear

1. Conditions of Operation   91
2. Speed and Torque Factors   91

8. Simple Planetary Gear Systems Employing Simple Gear Set Without Ring Gear

1. Numbers of Gear Teeth   94
2. Conditions of Operation   94
3. Condition 1   94
4. Condition 2   96
5. Conditions 3–4   96
6. Conditions 5–6   97
7. Conditions 7–12   97
8. Conditions 13–18   99
9. Summary and Graphs   99

9. Other Simple and Compound Planetary Gear Systems

1. Conditions of Operation 100
2. Speed and Torque Factors 100

10. Transmission Control Systems

1. General Description 104
2. Functions 105
3. Automatic Transmission Controls 109
4. Effects on Torque and Speed Factors 115

11. Model T Ford Transmissions

1. General Description 119
2. Transmission Controls 120
3. Numbers of Gear Teeth 120
4. Condition of Operation in Low and Reverse Gears 121
5. Low-Gear Power-On Operation 121
6. Reverse-Gear Power-On Operation 121
7. High-Gear Power-On Operation 121
8. Neutral-Gear Operation 121
9. Braking Operations 122
10. Power-Off Operation 122
11. Control System Considerations 123

12. TorqueFlite Transmissions

1. General Description 126
2. Numbers of Gear Teeth 127
3. 1st-Gear Power-On Operation (In Drive or Second Position) 127
4. 2nd-Gear Power-On Operation (In Drive Position) 128
5. 3rd-Gear Power-On Operation (In Drive Position) 129
6. Reverse-Gear Power-On Operation 129
7. Neutral-Gear Operation (In Neutral or Park Position) 129
8. First-Gear Power-On Operation (In First Position) 129
9. Second-Gear Power-On Operation (In Second Position) 130
10. 1st-Gear Power-Off Operation (In Drive or Second Position) 130
11. First-Gear Power-Off Operation (In First Position) 131
12. Second-Gear Power-Off Operation (In Second Position) 132
13. 2nd-Gear Power-Off Operation (In Drive Position) 132
14. 3rd-Gear Power-Off Operation (In Drive Position) 132
15. Reverse-Gear Power-Off Operation 133
16. Control System Considerations 133

13. Powerglide Transmissions

1. General Description 136

   2. Numbers of Gear Teeth   136
   3. 1st-Gear or Low-Gear Power-On Operation (In Drive or
       Low Position)   136
   4. High-Gear Power-On Operation (In Drive Position)   138
   5. Reverse-Gear Power-On Operation   138
   6. Neutral-Gear Operation (In Neutral or Park Position)   138
   7. Power-Off Operation   138
   8. Control System Considerations   138

14. Dual-Path Turbine Drive Transmissions

   1. General Description   140
   2. Numbers of Gear Teeth   140
   3. 1st-Gear Power-On Operation (In Drive Position)   142
   4. 1st-Gear Power-Off Operation (In Drive Position)   142
   5. High-Gear Power-On Operation (In Drive Position)   142
   6. High-Gear Power-Off Operation (In Drive Position)   145
   7. Low-Gear Power-On Operation (In Low Position)   145
   8. Low-Gear Power-Off Operation (In Low Position)   146
   9. Reverse-Gear Power-On Operation   146
  10. Reverse-Gear Power-Off Operation   146
  11. Neutral-Gear Operation (In Neutral or Park Position)   147
  12. Control System Considerations   147

15. Super Turbine "300" Transmissions

   1. General Description   149
   2. Numbers of Gear Teeth   150
   3. 1st-Gear or Low-Gear Power-On Operation (In Drive or
       Low Position)   151
   4. High-Gear Power-On Operation (In Drive Position)   151
   5. Neutral-Gear Operation (In Neutral or Park Position)   151
   6. Reverse-Gear Power-On Operation   151
   7. Power-Off Operation   151
   8. Control System Considerations   152

16. Super Turbine "400" Transmissions

   1. General Description   153
   2. Numbers of Gear Teeth   153
   3. 1st-Gear Power-On Operation (In Drive Position)   153
   4. 1st-Gear Power-Off Operation (In Drive Position)   155
   5. Low-Gear Power-On Operation (In Low Position)   155
   6. Low-Gear Power-Off Operation (In Low Position)   155
   7. 2nd-Gear or Second-Gear Power-On Operation (In Drive
       or Low Position)   156
   8. 2nd-Gear or Second-Gear Power-Off Operation (In Drive
       or Low Position)   156

9. 3rd-Gear Power-On Operation (In Drive Position)   156
10. 3rd-Gear Power-Off Operation (In Drive Position)   156
11. Reverse-Gear Power-On Operation   157
12. Reverse-Gear Power-Off Operation   157
13. Neutral-Gear Operation (In Neutral or Park Position)   157
14. Control System Considerations   157

17. Dynaflow Transmissions

1. General Description   160
2. Torque Converter and Associated Gear Set   160
3. Numbers of Gear Teeth   162
4. Drive-Gear Power-On Operation (In Drive Position)   162
5. Drive-Gear Power-Off Operation (In Drive Position)   165
6. Low-Gear Power-On Operation (In Low Position)   166
7. Low-Gear Power-Off Operation (In Low Position)   166
8. Reverse-Gear Power-On Operation   167
9. Reverse-Gear Power-Off Operation   167
10. Neutral-Gear Operation (In Neutral or Park Position)   167
11. Control System Considerations   167

18. Cruise-O-Matic and Merc-O-Matic Transmissions

1. General Description   169
2. Numbers of Gear Teeth   169
3. 1st-Gear Power-On Operation (In Drive 1 Position)   171
4. 1st-Gear Power-Off Operation (In Drive 1 Position)   171
5. Low-Gear Power-On Operation (In Low Position)   171
6. Low-Gear Power-Off Operation (In Low Position)   171
7. 2nd-Gear Power-On Operation (In Drive 1 or 2 Position)   172
8. 2nd-Gear Power-Off Operation (In Drive 1 or 2 Position)   172
9. 3rd-Gear Power-On Operation (In Drive 1 or 2 Position)   172
10. 3rd-Gear Power-Off Operation (In Drive 1 or 2 Position) 173
11. Reverse-Gear Power-On Operation   173
12. Reverse-Gear Power-Off Operation   173
13. Neutral-Gear Operation (In Neutral or Park Position)   173
14. Control System Considerations   174

19. Studebaker Automatic Transmissions

1. General Description   175
2. Numbers of Gear Teeth   175
3. 2nd-Gear Power-On Operation (In Drive Position)   177
4. 2nd-Gear Power-Off Operation (In Drive Position)   177
5. 2nd-Gear Hill-Holding Operation (In Drive Position)   177
6. 3rd-Gear Power-On Operation (In Drive Position)   178
7. 3rd-Gear Power-Off Operation (In Drive Position)   178
8. Low-Gear Power-On Operation (In Low Position)   178

    9. Low-Gear Power-Off Operation (In Low Position)   179
  10. Reverse-Gear Power-On Operation   179
  11. Reverse-Gear Power-Off Operation   180
  12. Neutral-Gear Power-On Operation (In Neutral or Park Position)   180
  13. Neutral-Gear Power-Off Operation (In Neutral Position)   181
  14. Control System Considerations   181

20. Hydra-Matic Transmissions

    1. General Description   183
    2. Numbers of Gear Teeth   183
    3. 1st-Gear Power-On Operation (In Drive Position)   185
    4. 1st-Gear Power-Off Operation (In Drive Position)   185
    5. 2nd-Gear Power-On Operation (In Drive or S Position)   185
    6. 2nd-Gear Power-Off Operation (In Drive or S Position)   187
    7. 3rd-Gear Power-On Operation (In Drive Position)   187
    8. 3rd-Gear Power-Off Operation (In Drive Position)   188
    9. 4th-Gear Power-On Operation (In Drive Position)   189
  10. 4th-Gear Power-Off Operation (In Drive Position)   189
  11. 1st-Gear Power-On Operation (In Low Position)   190
  12. 1st-Gear Power-Off Operation (In Low Position)   191
  13. Second-Gear Power-On Operation (In Low Position)   191
  14. Second-Gear Power-Off Operation (In Low Position)   191
  15. 1st-Gear Power-On Operation (In S Position)   191
  16. 1st-Gear Power-Off Operation (In S Position)   192
  17. Third-Gear Power-On Operation (In S Position)   192
  18. Third-Gear Power-Off Operation (In S Position)   192
  19. Reverse-Gear Power-On Operation   192
  20. Reverse-Gear Power-Off Operation   193
  21. Neutral-Gear Operation (In Neutral or Park Position)   193
  22. Control System Considerations   194

Appendix A: Definitions

    1. Directions of Rotation   197
    2. Gears   197
    3. Loads   198
    4. Planetary Gear Sets   198
    5. Planetary Gear Systems   198
    6. Speed and Torque Factors   199
    7. Power Conditions   200

Appendix B: Symbols

Appendix C: Nonplanetary Gear Equivalents of Planetary Gear Sets in Figures 11 and 17

1. Equivalent Planet Carrier Gear Teeth   203
2. Differential Action of Planetary Gear Sets   203
3. Nonplanetary Gear Equivalents of Planetary Gear Sets   206

List of Illustrations

1. Nonplanetary Gear Sets   2
2 A & B. Automobile Power and Torque Characteristics   a-10 b-11
3. Sliding-Gear Transmission and Clutch   19
4. Fluid Clutch   25
5. Fluid Clutch Torques   33
6. Fluid Clutch and Automobile Torques   35
7. Three-Element Torque Converter   40
8. Simulation of Three-Element Torque Converter   42
9. Torque Factors of Three-Element Torque Converter   47
10. Three-Element Torque Converter and Automobile Torques   50
11. Simple Planetary Gear Set with Ring Gear   56
12. Speed and Torque Factors—Form 1   63
13. Speed and Torque Factors—Forms 2 and 3   64
14. Compound System Output-Shaft Speeds-Vs.-Torque Requirements (Simple Gear Set—Fig. 11)   80
15. Compound System Condition 5 Speed-Vs.-Torque Relationships (Simple Gear Set—Fig. 11)   81
16. Compound System Condition 5 Speed Relationships (Simple Gear Set—Fig. 11)   82
17. Compound Planetary Gear Set with Ring Gear   84
18. Speed and Torque Factors—Simple System Conditions 11–14 and 19–20 (Compound Gear Set—Fig. 17)   89
19. Simple Planetary Gear Set without Ring Gear   95
20. Simple System Speed and Torque Factors (Simple Gear Set—Fig. 19)   98
21. Simple Planetary Gear Sets   102, 103
22. Engine and Automobile Speeds   106
23. Model T Ford Transmission   119
24. TorqueFlite Transmission   126
25. Powerglide Transmission   137
26. Dual-Path Turbine Drive Transmission   141

27. Dual-Path Turbine Drive Transmission in High Gear   143
28. Super Turbine "300" Transmission   150
29. Super Turbine "400" Transmission   154
30. Dynaflow Transmission   161
31. Cruise-O-Matic and Merc-O-Matic Transmissions   170
32. Studebaker Automatic Transmission   176
33. Hydra-Matic Transmission   184
34. Equivalents of Planetary Gear Set Shown in Fig. 11   204
35. Differential Gear Set   206
36. Nonplanetary Equivalents of Figs. 11 and 17   208
37. Modifications of Gear System Shown in Fig. 36   209

List of Tables

1. Speed Factors of Gear Systems Shown in Fig. 1 A   3
2. Speed Factors of Gear Systems Shown in Fig. 1 C   4
3. Transmission Torque Factors   20
4. Speed and Torque Factors in Conditions 1–6   62
5. Torque Relationships   72
6. Conditions 1–6—Fig. 21 D   100
7. Transmission Shift Pattern   118
8. Transmission Control Positions   120
9. Power-Off Speed and Torque Factors   123
10. Control System Pattern—TorqueFlite   134
11. Control System Pattern—Dual-Path Turbine Drive Transmissions   147
12. Control System Pattern—Super Turbine "400" Transmissions   158
13. Control System Pattern—Cruise-O-Matic and Merc-O-Matic Transmissions   174
14. Control System Pattern—Studebaker Automatic Transmissions   182
15. Control System Pattern—Hydra-Matic Transmissions   194

Charts

1. Simple Planetary Gear Systems (Simple Gear Set—Fig. 11)   63
2. Applicability of Figs. 12 and 13   65
3. Compound Planetary Gear Systems (Simple Gear Set—Fig. 11)   81
4. Simple Planetary Gear Systems (Compound Gear Set—Fig. 17)   90
5. Simple Planetary Gear Systems (Simple Gear Set—Fig. 19)   99

AUTOMATIC TRANSMISSIONS

1

Introduction

1. Principles of Simple Gear Systems

Figure 1 illustrates gear sets in which the gear shafts rotate with their gears, but otherwise are stationary. Let the input shaft of a gear system employing the gears in Fig. 1 A be the shaft of gear 1, and S_1 be the speed in revolutions per minute (rpm) of gear 1 and its shaft. When gear 1 turns one revolution, each tooth on gear 2 moves through an angle which includes T_1 teeth of gear 2, or through T_1/T_2 revolutions. Since the revolution of gear 1 and the T_1/T_2 revolutions of gear 2 occur in the same time period, when gear 1 has a speed of S_1 rpm the speed of gear 2 and its shaft in rpm, without consideration of the direction of rotation, is

$$S_2 = S_1\,(T_1/T_2) = S_{in}\,(T_1/T_2),\ \text{or}$$
$$S_{out} = S_{in}\,(T_1/T_2)$$

if the shaft of gear 2 is the output shaft of the gear system. The speed and the direction of rotation of each of the gears and shafts are indicated in the following equation when each shaft, in turn, is considered to be the output shaft. Appendices A and B include definitions of terms and meanings of symbols, including definitions of the forward and reverse directions.

$$S_{out} = (+S_{in})\,(-T_1/T_2)\,(-T_2/T_3)\,(-T_3/T_4)\,(-T_4/T_5)$$

Shaft 1
Shaft 2
Shaft 3
Shaft 4
Shaft 5

In this equation, the plus $(+)$ sign indicates that the input shaft rotates in the forward direction, a minus $(-)$ sign that the direction of rotation of a gear and shaft is the opposite of that of the gear and shaft preceding

1

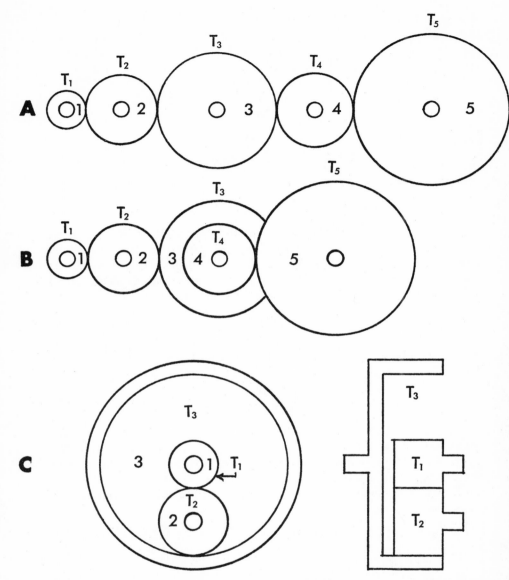

FIG. 1. Nonplanetary Gear Sets. (1) T_1 etc., are numbers of gear teeth on gear 1, etc. (2) Each shaft rotates with its gear, but otherwise is stationary. (3) Compound gear in Fig. 1 B is one piece of metal.

it. Multiplication of the terms pertaining to each of shafts 2 to 5 results in the values shown in Table 1. The *In* column lists the input shaft and the *Out* column identifies the shaft that is considered to be the output shaft insofar as the entries on its line of the listing are concerned.

TABLE 1. SPEED FACTORS OF GEAR SYSTEMS SHOWN IN FIG. 1 A

Shafts		Gears and Shafts Used		S_{out}		S_{out}/S_{in}, or Speed Factor (SF)
In	Out	No.	Even or Odd	Value	Direction	
1	2	2	Even	$S_{in}\,(-T_1/T_2)$	Reverse	$(-T_1/T_2)$
1	3	3	Odd	$S_{in}\,(+T_1/T_3)$	Forward	$(+T_1/T_3)$
1	4	4	Even	$S_{in}\,(-T_1/T_4)$	Reverse	$(-T_1/T_4)$
1	5	5	Odd	$S_{in}\,(+T_1/T_5)$	Forward	$(+T_1/T_5)$

It is evident from Table 1 that the values of the output shaft speed and the speed factor of this type of gear system, disregarding the signs of the values, are independent of the number of gears and shafts employed between the input gear and shaft and the output gear and shaft, and independent of the numbers of teeth on the intermediate gears. In each case, disregarding the sign, the speed factor is equal to T_{in}/T_{out}, where T_{in} and T_{out} are the numbers of teeth on the input and output gears. The signs of the output speed and the speed factor are determined by the number of intermediate gears and shafts, or by the total number of gears and shafts in the system, including the input and output gears and shafts. The signs are $+$ if the total number is an odd number, and $-$ if it is an even number.

Figure 1 B shows a gear system in which one of the shafts has a compound gear secured to it. The speed factor of this gear train when shaft 1 is the input shaft is

$$SF = S_{out}/S_{in} = (-T_1/T_2)\,(-T_2/T_3)\,(+1)\,(-T_4/T_5)$$
$$= (-T_1/T_5)\,(T_4/T_3)$$

In this case, the speed factor of the gear train, disregarding the sign, is equal to the ratio of T_{in}/T_{out} of the train, multiplied by the ratio of T_{out}/T_{in} of the compound gear. The sign of the speed factor is minus, the number of gears is even (considering the compound gear to be one gear), and the number of shafts is even. Studies of other gear trains like that shown in Fig. 1 B, but in which there are more than one compound gear, show that the value of the speed factor, disregarding the sign, is

$$SF = (T_{in}/T_{out}\ \text{of train})\,(T_{out}/T_{in}\ \text{of compound gear 1})$$
$$(T_{out}/T_{in}\ \text{of compound gear 2})$$

3

(and so on for other compound gears). The sign is $+$ if the number of shafts and gears is an odd number, and $-$ if this number is even, when each of the compound gears is considered as being one gear.

Figure 1 C shows a gear system which resembles that depicted in part A of Fig. 1, except that one of the gears is similar to the ring gear, or internal gear, of a planetary gear set (see Fig. 11). Though the system has the appearance of a planetary gear set, it is not one since gear 2 and its shaft cannot rotate around the axis of gear 1 as they could if the system were a planetary set.

Let the various shafts in Fig. 1 C be input and output shafts as indicated in Table 2. The shaft which is not listed in any particular line is Free (to rotate). In this case, the relationship between the sign of the speed factor and the nature of the number of gears and shafts used in the system is not a constant relationship, as it is in the preceding systems.

TABLE 2. SPEED FACTORS OF GEAR SYSTEMS SHOWN IN FIG. 1 C

Shafts		No. of Gears and Shafts Used	Speed Factor
In	Out		
1	3	Odd	$-T_1/T_3$
3	1	Odd	$-T_3/T_1$
2	3	Even	$+T_2/T_3$
3	2	Even	$+T_3/T_2$
1	2	Even	$-T_1/T_2$
2	1	Even	$-T_2/T_1$

A gear may be considered to be a special form of lever having a fulcrum at the axis of the shaft of the gear. The end(s) of the lever is (are) at the junction(s) of the teeth of the gear and those of the adjacent gear(s). A force may be transferred from one gear to another because of the overlapping of the ends of the levers formed by the two gears. The distances from the fulcrum to the ends of a lever of one gear are equal, so the force at one end of the lever is equal in magnitude to that at the other end. Therefore, a force of 1 pound (lb) which is placed in the downward direction on a tooth of gear 1 in Fig. 1 A at the left extremity of that illustration produces a downward force of 1 lb on a tooth of gear 5 at the junction of gears 4 and 5. Thus, the gear train does not increase or decrease the applied force but, as shown below, may increase or decrease the torque that is produced by the applied force.

Torque is the product of a force and the length of the torque arm that is used in transmitting the force to a shaft or other object. For example, a force of 1 lb on the handle of a crank which has a radius of 1 foot (ft) and is attached to the shaft of any of the gears of Fig. 1 produces a torque of 1 pound-foot (lb-ft) on that shaft when the force is applied at each instant in a direction which is perpendicular to a straight line connecting the crank handle and the axis of rotation. A force of 2 lb on the crank handle produces a torque of 2 lb-ft. Likewise, a force of 2 lb on a crank which has a radius of 2 ft produces a torque of 4 lb-ft.

Torque may be increased or reduced by a gear system, or it may be unchanged. For example, assume that:

(1) A 1 lb force is placed on a 1 ft crank connected to the shaft of gear 1 in Fig. 1 A in a direction which is at all times perpendicular to a straight line connecting the crank handle and the axis of the shaft of gear 1,

(2) The radii of gears 1 and 2 are 1 and 2 ft,

(3) T_1 and T_2 are 100 and 200 teeth.

The 1 lb force produces a torque of 1 lb-ft on the shaft of gear 1. It also produces a force of 1 lb on a tooth of gear 2, and this force produces a torque on the shaft of gear 2 of 2 lb-ft if gears 3 to 5 can rotate freely, since the radius of gear 2 is 2 ft. However, this torque is in the direction opposite to that on the shaft of gear 1. Thus, the torque transmitted by the shaft of gear 2 to a load connected to that shaft is

$$Tq_{\text{out shaft 2}} = (Tq_{\text{in shaft 1}}) (R_{\text{gear 2}}/R_{\text{gear 1}}) (-1)$$
$$= (Tq_{\text{in shaft 1}}) (-2)$$
$$= (Tq_{\text{in shaft 1}}) (-T_2/T_1)$$

where the — signs indicate the reverse direction with respect to the direction of $Tq_{\text{in shaft 1}}$.

The torque factor of this two-gear gear train is

$$TF = Tq_{\text{out}}/Tq_{\text{in}} = -T_2/T_1$$

The output torque of the entire gear train is

$$Tq_{\text{out shaft 5}} = (Tq_{\text{in}}) (-T_2/T_1) (-T_3/T_2) (-T_4/T_3) (-T_5/T_4)$$
$$= (Tq_{\text{in}}) (T_5/T_1)$$

The torque factor of the gear train is

$$TF = Tq_{\text{out}}/Tq_{\text{in}} = T_5/T_1$$

Comparison of the values in the above series with the corresponding values in the series pertaining to the speeds within the gear train in Fig. 1 A, as stated in the first part of this section, shows that the torque factor in each instance is equal to $1/SF$. This relationship is true in any simple gear system capable of transmitting torque—that is, any system which has only one input shaft and one output shaft and transmits

5

torque from one to the other. If conditions in the gear system are such that it cannot transmit torque,

$$Tq_{out}/Tq_{in} = TF = 0.0$$

The output shaft speed, then, is zero, and

$$SF = S_{out}/S_{in} = 0.0$$

The torque applied to a shaft or a gear at any instant is equal in magnitude to the torque the load (directly or indirectly) connected to the shaft or gear imposes on it at that instant, including the torque the load may impose because of its acceleration or deceleration, but the two torques are in opposite directions. This statement disregards the torque involved in accelerating or decelerating the shaft or gear and such parts as are attached to it if the speed is changing, but normally this torque is small in comparison with the torque applied to the shaft or gear. In constant-speed operations, the torque imposed on the output shaft by the load is equal to the torque required to drive the load at its speed, but the two torques have opposite signs.

If the speed factor of a simple gear system has a value of N when shaft X is used as an input shaft and shaft Y is used as an output shaft, then the speed factor is $1/N$ when the functions of shafts X and Y are exchanged. The same relationship applies to the torque factors of a simple gear system.

The power applied to an input shaft is $(Tq_{in})(S_{in})(2\pi)$, as stated in §3, below. Likewise, the power provided by an output shaft to its load is $(Tq_{out})(S_{out})(2\pi)$. Assuming that the power losses within the gear system due to friction are negligible,

$$\text{Input Power} = \text{Output Power, or}$$
$$(Tq_{in})(S_{in})(2\pi) = (Tq_{out})(S_{out})(2\pi), \text{ or}$$
$$Tq_{out}/Tq_{in} = S_{in}/S_{out}$$

Thus, increases in torque multiplication by gear systems are purchased at the cost of increases in the ratio S_{in}/S_{out}. In effect, increases in output torques are exchanged for decreases in output speeds, and increases in output speeds are exchanged for decreases in output torques, when the input power remains constant. In the case of automobiles which are being started in motion, the engine is capable of much greater speed than is required by the automobile driveshaft, but the engine torque may be insufficient to turn the driveshaft if the engine is coupled directly to the driveshaft. The use of gear sets between the engine and the driveshaft permits an exchange of high engine speed for high driveshaft torque.

2. Speed and Torque Factors

Gear systems are employed in automobile transmissions for the same reasons that cause them to be used in other applications of gear systems. These include the providing of a means of matching the speed and torque capabilities of a source of power, and the speed and torque requirements of a load which permits a function of the mechanism in which the gear system is used to be served in an effective manner. In some cases, the relative speeds of the input and output shafts of the gear system are of primary importance; in others, the relative torques are of primary importance; in still others, both are important. Thus, the speed and torque characteristics of gear systems are of basic interest in a study of transmissions. The relationships between the speed and torque characteristics of power sources and loads are discussed in a following section.

One of the speed characteristics of a simple gear system—a system which has only one input shaft and one output shaft—may be stated in terms of its speed factor, which, as indicated in § 1, is

$$SF = S_{out}/S_{in} = 1/TF$$

Compound planetary gear systems of the types discussed in later chapters have two input shafts and one output shaft, or one input shaft and two output shafts. Therefore, the above definition of the speed factor is inadequate in these cases.

When a system employing the gear set shown in Fig. 11 has two input shafts and one output shaft, the speeds of the input shafts may have any relationship to one another, but this relationship may be stated as

$$\pm S_{in\ 2}/S_{in\ 1} = A$$

where the direction of rotation of input shaft 1 is considered to be the forward direction, the $+$ sign is used if the rotation of input shaft 2 is in the same direction as that of input shaft 1, and the $-$ sign is used if it is in the opposite direction. Then, for a specified value of A, the speed factors of the system are expressed as

$$SF_1 = S_{out}/S_{in\ 1}, \text{ and}$$
$$SF_2 = S_{out}/S_{in\ 2}$$

The factor A is always negative in certain Conditions of Operation (see Appendix A, Definitions) and always positive in another—that is, when certain shafts are used as input shafts and the third shaft is used as the output shaft.

When a system has one input shaft and two output shafts, the speed factors are

$$SF_1 = S_{out\ 1}/S_{in}, \text{ and}$$
$$SF_2 = S_{out\ 2}/S_{in}$$

7

for a specified value of B where $B = S_{out\ 2}/S_{out\ 1}$ and the direction of rotation of the input shaft is considered as being the forward direction. The factor B is always negative in certain Conditions of Operation and always positive in another.

In a simple gear system, the torque factor is

$$TF = Tq_{out}/Tq_{in} = 1/SF$$

As in the case of the speed factors, there are two torque factors which apply to compound gear systems employing the gear set shown in Fig. 11.

When a system has one input shaft and two output shafts, the torque factors are

$$TF_1 = Tq_{out\ 1}/Tq_{in}, \text{ and}$$
$$TF_2 = Tq_{out\ 2}/Tq_{in}$$

When there are two input shafts and one output shaft,

$$TF_1 = Tq_{out}/Tq_{in\ 1}, \text{ and}$$
$$TF_2 = Tq_{out}/Tq_{in\ 2}$$

Let it be assumed that the speed factor, SF_1, of a particular compound system which employs the gear set shown in Fig. 11 and has two input shafts and one output shaft is

$$SF_1 = S_{out}/S_{in\ 1} = X$$

where X is a particular numerical value. From the definition of the factor A,

$$S_{in\ 2} = A\ S_{in\ 1}$$

where A is a positive or negative value as indicated above. Then,

$$SF_2 = S_{out}/S_{in\ 2} = S_{out}/(A\ S_{in\ 1}), \text{ and}$$
$$SF_2 = SF_1/A = X/A$$

When a particular compound system has one input shaft and two output shafts, computations such as the above show that

$$SF_2 = B\ SF_1 = B\ X$$

where X is the particular value of SF_1 of the system and B is a positive or negative value as indicated above.

When a particular compound system employing the above gear set has one input shaft and two output shafts, the torque factors are not dependent on any variable factor such as A or B, so

$$TF_2 = C\ TF_1$$

where C is a constant having a magnitude determined by the numbers of teeth on the gears of the gear set. It is a positive value when B is positive and negative when B is negative.

8

When a particular compound system employing the gear set shown in Fig. 11 has two input shafts and one output shaft, computations such as the above show that

$$TF_2 = TF_1/U = X/U$$

where X is the value of TF_1 of the particular system and

$$U = \pm Tq_{in\ 2}/Tq_{in\ 1}$$

The value of U is a constant which has a magnitude determined by the numbers of teeth on the gears of the gear set. It is negative in certain Conditions of Operation, positive in another.

The relationships between SF_1 and SF_2 and between TF_1 and TF_2 in compound systems employing gear sets other than those illustrated in Fig. 11 may be the same as or similar to the relationships above, or they may differ with them, as indicated in Chapter 7.

3. Work, Power, Energy, Etc.

The units of measurement most used herein are the pound, foot, minute, second, revolution, radian, horsepower, and power in foot-pounds per second or per minute. A radian is an angle at the center of a circle which has an arc that is equal to the radius of the circle. Therefore, there are 2π radians in a revolution, or approximately 57 degrees in a radian.

Work may be defined as the movement of an object by a force, or

Work = Distance an object is moved × the Force required to move the object, or

Work = $D\ F$, or

Work = (Feet) (Pounds) = ft-lb

Torque, discussed in §2, is the product of a force and the length of the torque arm. The length of the torque arm is the radius at which the force is applied when measured from the axis of rotation. Then,

Tq = (Force) (Length)
   = (Pounds) (Feet) = lb-ft

The work done by a torque is:

Work = (F) (Length of torque arm)(2π)(No. of revolutions of torque arm), or

Work = $(Tq)(2\pi)$(No. of revolutions), or

Work = $(Tq)(2\pi)$(Speed)(Time), where the speed is in revolutions per minute (rpm) and Time is the number of minutes during which the revolutions occur, or

Work = (Tq)(No. of radians through which the torque arm is moved), or

9

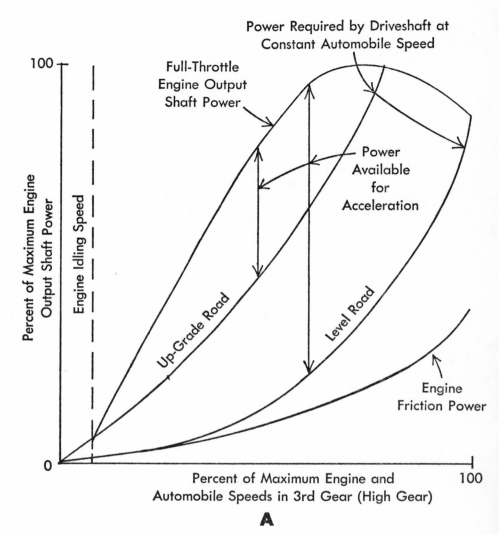

FIG. 2 A. Automobile power and torque characteristics.

Work $= (Tq)(W)(t)$, where W is the velocity of rotation of the torque arm in radians per second, and t is the time in seconds during which the rotation occurs. (The velocity in radians normally is expressed in radians per second, rather than radians per minute.)

Power may be defined as the amount of work that is done per unit of time, or

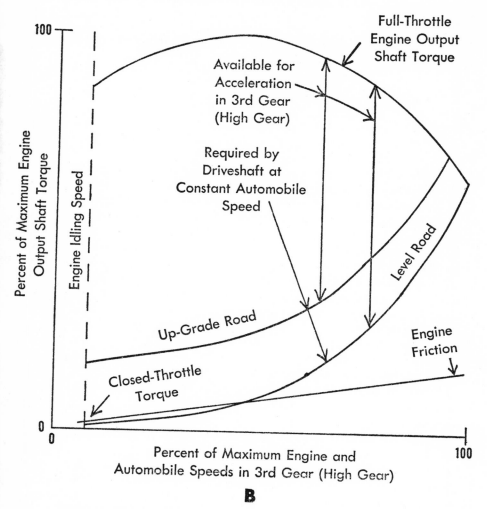

B

Fig. 2 B. Automobile power and torque characteristics (*Continued*).

Power = Work/Time
     = ft-lb/min, or ft-lb/sec

Then,

Power = $D\,F$/min, or $D\,F$/sec, or
Power = $(Tq)(2\pi)(S)$ ft-lb per minute, where S is the speed of
     rotation of the torque arm in revolutions per minute, or
Power = $(Tq)(W)$ ft-lb per second

11

The unit of power usually used when discussing automobiles is the horsepower. It is

$$1 \text{ horsepower} = 33,000 \text{ ft-lb of work per minute}$$

Thus, the horsepower provided by an engine to its load is

$$\begin{aligned}
\text{Horsepower} &= (\text{Work done per minute in ft-lb})/33,000, \text{ or}\\
&= (Tq)(S)(2_\pi)/33,000, \text{ or}\\
&= (Tq)(S)/5,252, \text{ where S is the speed of the engine}\\
&\quad \text{in rpm and } Tq \text{ is its output torque in lb-ft}
\end{aligned}$$

Figure 2 illustrates the torque and power capabilities of an automobile engine versus speed. Also included in the illustration are the power and torque required to drive the automobile versus speed. The power and torque required by the engine to drive itself are indicated as friction horsepower and friction torque since this horsepower and torque are used in overcoming the effects of various types of friction in the engine, including pumping friction (which is the friction involved in pumping the fuel-air and exhaust gas mixtures into and out of the engine cylinders). For the purposes of this discussion, friction power and torque include those used in driving the cooling fan, generator, power-steering fluid pump, etc., although some of these more properly are classified as output power and torque. The friction power and torque when added to the engine output shaft power and torque are the total mechanical power and total torque developed within the engine.

When an automobile is operating in a power-off condition, such as when decelerating on a level road or traveling down a steep hill with the engine throttled, power is transmitted by the automobile to the engine if the transmission transmits torque to the engine and the engine crankshaft is rotating. The amount of power transmitted to the engine is the friction horsepower of the engine at the speed at which the engine crankshaft is being driven if the ignition is off. If the ignition is on, the amount of power transmitted is this same amount less the friction horsepower at the speed the engine would have if it were disconnected from the transmission. This speed is the engine idling speed if the throttle is closed. If torque is transmitted to the engine, the amount transmitted is the amount required to drive the engine at its speed less the amount the engine would provide if it were disconnected from the transmission, as in the case of the power that is transmitted to the engine. (These statements concerning power and torque transmitted to the engine disregard the effects of certain factors which may cause the statements to be in error by small amounts.)

The energy of a body is defined as its capacity to do work. It may exist in any one or more of many forms, such as thermal and chemical energy, but two forms of particular interest in the study of transmissions and their use in automobiles are potential energy and kinetic energy.

The potential energy of a body is the energy the body possesses because of its position. For example, an automobile which weighs 3,000 lb and is on the top of a hill having an altitude of 1,000 ft above sea level is capable of doing 3,000,000 ft-lb of work in descending from the top of the hill to sea level. Therefore, it has 3,000,000 ft-lb of potential energy.

Kinetic energy is the energy a body possesses because of its motion. It is

$$\text{Kinetic energy} = (M/2) \, V^2$$

where the energy is in ft-lb, M is the mass of the body (which is its weight in pounds divided by 32.2), and V is the linear velocity of the body in feet per second.

Energy of one form may be converted into energy of another form, but energy cannot be destroyed. In many if not all instances, some of the energy in a system is converted to a form which is not useful insofar as the system is concerned. In the case of transmissions, the nonuseful energy is thermal energy, which results from friction and is dissipated in the form of heat. The extent to which the work, power, or energy that enters a system is available in useful form at the output of the system often is termed the efficiency of the system. The efficiency always is less than 1.0, or 100 percent, but in many instances it may be considered to be 1.0, or 100 percent, with a small error. When so considered, work input = work output, or power input = power output, or energy input = energy output. In other cases,

$$\text{Work output} = (\text{Eff.})(\text{Work input}), \text{ or}$$
$$\text{Power output} = (\text{Eff.})(\text{Power input}), \text{ or}$$
$$\text{Energy output} = (\text{Eff.})(\text{Energy input}),$$

where Eff. is the work, power, or energy efficiency, and the output in each case is the useful output. Thus, in general, Eff. = Output/Input when expressed as a fraction, or Eff. = (100%)(Output)/Input when expressed as a percentage.

Well-designed and -lubricated gear systems have high efficiencies, so the efficiency may be considered to be 1.0 from the standpoint of computing torque factors of gear systems with negligible error. The assumption that friction forces are negligible and the efficiency is 1.0 is used in succeeding chapters in which torque factors of gear systems are computed.

Acceleration of a body is the change in velocity of the body per unit of time, or

$$a = (V_1 - V_0)/t$$

13

where a is the acceleration in feet per second per second, V_1 and V_0 are the velocities of the body at the end and beginning of the time period in feet per second, and t is the length of the time period in seconds. Acceleration may be positive, which is its sign in the frequently used meaning of the word, acceleration, or it may be negative, in which case the word deceleration frequently is used to describe it. As stated above, acceleration is expressed in feet per second per second. Since velocity is expressed in feet per second, acceleration is the change in velocity of a body per second. It may or may not be a constant value during a second or longer or shorter period of time. In fluid clutches and torque converters, the acceleration of the fluid, in general, varies from one small fraction of a second to the next.

The relationship between linear acceleration and the force causing the acceleration is

$$F = M a = M(V_1 - V_0)/t, \text{ or}$$
$$F t = M(V_1 - V_0)$$

where F is in pounds and the other values are as in the above.

When a very small body rotates around an axis of rotation at a linear velocity which is constant in magnitude and in a circular path, the centrifugal force in pounds exerted by the body upon the container of the body or other device (such as a string) is

$$F = M a = M V^2/R$$

where R is the radius of the circular path in feet and the other values are as in the above. The acceleration in this case is the continuous change in the direction of the velocity of the body toward the center of the circular path. The magnitude of V is equal to $2\pi R S$, where S is the rotational speed in revolutions per second. When radians are used instead of revolutions, V is equal to $W R$ where W is the rotational or angular velocity of the object in radians per second and R is the radius in feet.

When the above equations are used in connection with the acceleration of an object larger than a very small object, the force, velocity and mass are considered to be concentrated at a point known as the center of gravity of the object. Then, the radius in the equation is the radius of the center of gravity from the axis of rotation, and V and a are the velocity and the acceleration of the center of gravity. Then the equation pertaining to circular motion becomes

$$F = M a = M V_g^2/R_g$$

where V_g is the velocity of the center of gravity and R_g is the radius of the center of gravity from the axis of rotation.

14

The equation pertaining to the torque required to cause an angular acceleration of a very small object is

$$Tq\,t = M\,R^2\,(W_a - W_b)$$

where the torque is in lb-ft, the radius is in feet, and W_a and W_b are the velocities in radians per second at the end and beginning of the time, t, in seconds.

When a larger object is considered, the product of the mass of each very small part of the object and the square of its radius must be used. However, the entire mass of the larger object may be considered to be located at a point which is known as the center of inertia of the object. The radius of the center of inertia from the axis of rotation is known as the radius of gyration. Then, for any object,

$$Tq\,t = M\,R_i^2\,(W_a - W_b)$$

where M is the mass of the object and R_i is the radius of gyration of the object.

The velocity of a liquid fluid which flows through a passageway such as a pipe of constant or varying diameter is

$$V = K\,\sqrt{2\,P/m} = K\,(2\,P/m)^{1/2}$$

where V is the velocity of the fluid in feet per second, m is the mass of the fluid per cubic foot, P is the difference in pounds per square foot between the pressures acting on the fluid at the ends of the passageway or at two places along the passageway, and K is a factor which is determined by items such as the viscosity of the fluid and the shape of the passageway.

2

Automatic Transmissions

1. Functions of Transmissions

As stated in Chapter 1, a general function of gear sets, including those used in transmissions, is to provide a matching of the speed and torque capabilities of a power source and the speed and torque requirements of a load which serves the intended purposes of the mechanism in which the gear set (or sets) is used.

Assume that a stationary automobile with characteristics such as those shown in Fig. 2 is to be placed in motion in the forward direction, and that the only device in the power train between the engine and the driveshaft is a clutch of a mechanical type. If an effort is made to start the automobile on level ground with the throttle closed or almost closed, the engine is apt to "die" because it has little if any torque in excess of that required to balance the torque of friction within it. If the same effort is made but on a steep up-grade road, the engine will "die" since the torque requirement of the automobile is considerably greater than the torque available from the engine. The automobile can be started in some instances if the engine is operated with a throttle opening such that the engine can provide an excess of torque and if the clutch is "slipped" until the automobile has a speed corresponding approximately with the engine speed. However, the clutch would be worn to an unserviceable state if many starts were made in this fashion. Thus, one function of a transmission is to provide a means by which the engine may be operated at relatively high speeds when the load (the automobile) operates at low speeds in order that the engine can provide greater power and torque for use in accelerating the automobile and/or overcoming the effects of up-grade roads or other conditions that increase the torque requirement. A related function is to multiply the torque of the engine and thus provide greater torque for such purposes.

Conversely, another function is to provide a means by which the engine can operate at relatively low speeds when the automobile speeds are high but the torque and power required by the automobile are well within the capabilities of the engine at its relatively low speeds.

Another function is to provide a means by which the engine may be used effectively in decelerating the automobile, or perform the function of a brake in maintaining reasonable automobile speeds on down-grade roads. An engine connected to the driveshaft by a mechanical clutch, only, would receive torque from the automobile through the driveshaft under these operating conditions. The speed of the engine at any given automobile speed would be the same as that when the engine drives the automobile. The power and torque delivered to the engine by the driveshaft are discussed in Chapter 1, § 3. If a gear train which has a speed factor greater than 1.0 (when the driveshaft is the power-source shaft) is employed in addition to the clutch, the ratio of engine speed to driveshaft speed (or automobile speed) is increased. The power and torque required to drive the engine at the greater speeds is provided by the automobile, and the providing of these causes a braking action on the speed of the automobile which is greater than it would be if a clutch only were used.

The above three functions can be served by a transmission which has two speed factors under power-on conditions, but three or four frequently are provided. For example, assume that the transmission provides either one of two power-on speed factors, 0.4 and 1.0. The power-on torque factors are $1/SF$, or 2.5 and 1.0. In starting the automobile, the 0.4 speed factor and the 2.5 torque factor are used. The engine speed for any given automobile speed is 2.5 times as great as it would be if a clutch only were used in the power train. The torque and power capabilities of the engine are increased by the amounts indicated in Fig. 2, and the torque provided by the engine is multiplied by a factor of 2.5. When the automobile speed has reached a suitable value, the gears are "shifted" so that the speed and torque factors are 1.0. The engine crankshaft and the driveshaft then rotate at the same speed and the ratio of engine speed to automobile speed is less than before the shift of gears occurred. When the automobile is descending a steep hill, the gears may be shifted to the condition used in starting the automobile. In this instance, the "new" input shaft of the transmission is the shaft that is connected to the driveshaft and the "new" output shaft is the shaft connected to the clutch and, through the clutch, to the engine. This exchange of functions of the shafts causes the speed and torque factors of the transmission to be 2.5 and 0.4, respectively, as indicated in Chapter 1, § 1. Thus, the engine speed is 2.5 times the driveshaft speed and the torque available to drive the engine is 0.4 times the torque provided by the driveshaft. If the torque available to drive the engine is less than that required to drive it at its speed, the speed of the driveshaft must decrease until the torque applied to the engine crank-

17

shaft is that which is required at the speed of the crankshaft. As the driveshaft speed decreases, the speed of the automobile decreases. In this manner, engine friction is used for braking purposes, and it is more effective than it would be if the gears were not shifted from the position in which the speed and torque factors are 1.0.

Another function of a transmission is to provide a means by which the engine can be disconnected from the driveshaft so that the engine crankshaft is free to rotate when the automobile is stationary. This function can be performed by a mechanical clutch operated by the driver, and is so performed in many instances. However, the use of a mechanical clutch for this function normally requires continuous pressure on the clutch pedal by the driver. A means of keeping the clutch disengaged without this action by the driver can be used. If a mechanism such as a ratchet is used, accidental release of the mechanism or failure of some part of it permits the clutch spring to engage the clutch and may start the automobile in motion with no driver in the automobile. For reasons such as this, the function of disconnecting the engine from the driveshaft is performed by the transmission in manually controlled transmission automobiles and by the gear portion of the transmission in automatic transmission automobiles.

Another function is to provide a means by which the automobile may be driven backward, or in a reverse direction. This transmission function is needed since the engines normally used in automobiles are capable of operating in only one direction of rotation.

Still another function can be and often is performed by the transmission. It is that of maintaining the driveshaft in a zero-speed condition, and thereby serving as a parking brake of the automobile. In manually controlled transmission automobiles, this usually is accomplished by placing the gear train in a condition such that its SF is greater than 1.0 and its TF is less than 1.0 when the shaft connected to the driveshaft is the input shaft of the transmission. Conditions then exist which are similar to those when the engine is used as a brake in descending a hill—that is, if the automobile moves it must drive the engine at a relatively high speed. In this instance, the ignition is off, whereas in descending a hill it usually is on. Therefore, the braking effect at any given automobile speed is greater than it normally is in downhill operations. The Low-gear or Reverse-gear position of the transmission control lever (gear-shift lever) frequently is used when the parking brake function is to be performed. In automatic transmissions a Park position of the transmission control device usually is provided. When this position is used, the conditions of the transmission gears, etc., are the same as when in the Neutral position, but the transmission output shaft is locked to the transmission case in such manner that it cannot rotate. Therefore, the driveshaft cannot rotate and the automobile is held stationary. The

18

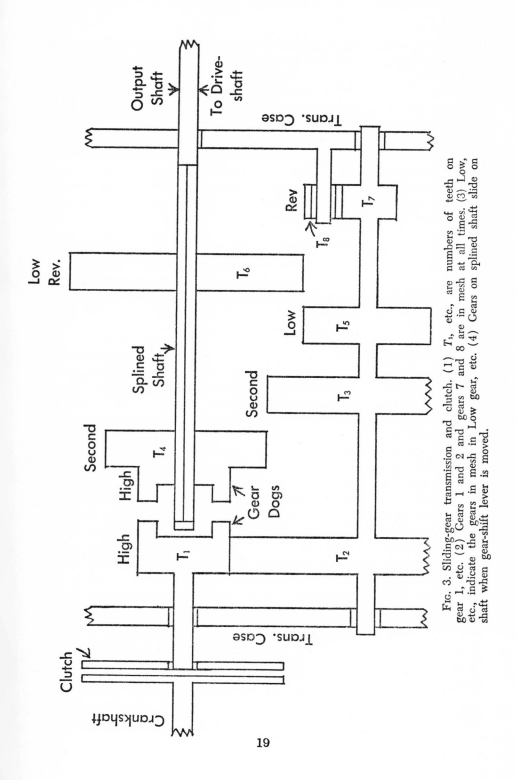

FIG. 3. Sliding-gear transmission and clutch. (1) T_1, etc., are numbers of teeth on gear 1, etc. (2) Gears 1 and 2 and gears 7 and 8 are in mesh at all times. (3) Low, etc., indicate the gears in mesh in Low gear, etc. (4) Gears on splined shaft slide on shaft when gear-shift lever is moved.

parking brake function of an automatic transmission can be performed when the engine is running, whereas that of the manually controlled transmission can be performed only when the engine is not running. Conversely, the placing of an automatic transmission in the Low, Reverse, or other Drive position does not hold the automobile stationary when the engine is not running since the transmission remains in Neutral gear due to a lack of the transmission fluid pressure which is needed to place it in another gear. Also, the fluid clutch or torque converter would transmit very little torque to the engine even though the gear portion of the transmission were "in gear." An exception to this would be the transmission shown in Fig. 32 if it could be in 3rd gear when the engine is not running since the torque converter is not used in 3rd gear.

2. Manually Controlled Transmissions

Figure 3 illustrates a manually controlled transmission of a sliding-gear type. Sliding gears were used throughout the transmissions of many automobiles until the Syncro-Mesh transmissions appeared in about 1930, and are used to provide some of the gear ratios in many Syncro-Mesh transmissions.

In placing the automobile in motion in the forward direction, the clutch is disengaged (as shown in Fig. 3) and the transmission control lever (gear-shift lever) is moved so that the Low gears are in mesh. Next, the engine is accelerated to a suitable speed and the clutch is engaged. After a suitable automobile speed is attained, the clutch is disengaged, the gear-shift lever is moved so that the Second, or Intermediate, gears are in mesh and the Low gears are not in mesh, and the clutch is engaged. Again, after suitable automobile speed is attained, the gear-shifting process is repeated so that the dogs of gear 1 and gear 4 are engaged. This process is reversed in shifting from High to Second and to Low gear when such shifting is used. Starting the automobile in the reverse direction is accomplished in the same manner as starting in Low gear except that the gear-shift lever is placed in the Reverse position.

The torque factors of the transmission when the engine is providing power to the transmission are shown in Table 3.

TABLE 3. TRANSMISSION TORQUE FACTORS

Gear	$TF = Tq_{out}/Tq_{in}$
Low	$(-T_2/T_1)(-T_6/T_5)$
Second	$(-T_2/T_1)(-T_4/T_3)$
High	1.0
Reverse	$(-T_2/T_1)(-T_8/T_7)(-T_6/T_8) = (-T_2/T_1)(T_6/T_7)$
Neutral	$(-T_2/T_1)(0.0) = 0.0$

The speed factor in any of the gears under the above condition is S_{out}/S_{in} and is equal to $1/TF$ of that gear, except in Neutral gear when the SF is 0.0.

When the automobile drives the output shaft (the "new" input shaft), the TF values are $1/TF$ as listed above and the SF values are equal to those listed except in Neutral gear when the TF and SF are 0.0.

Another manually controlled transmission is illustrated in Fig. 23. It employs a planetary gear set.

3. Automatic Transmission Components

The major components of automatic transmissions may include:

(1) A fluid clutch or fluid torque converter, usually but not always placed in the power train between the engine and the gears of the transmission. In at least one instance two fluid clutches are used.

(2) One or more planetary gear sets.

(3) Clutches or other devices which cause the gear sets to be in one or another of the different possible Conditions of Operation, and therefore to have one or another of the possible sets of speed and torque factors, in accordance with the actions of the transmission control devices in (4), below. In some instances, these components affect the operation of items other than the gear sets such as the torque converter.

(4) Transmission control devices including those which are operated by the driver and those which operate automatically. These may include a transmission control lever (gear-shift lever), the engine throttle control (accelerator) which serves also as a transmission control, electric circuits and components, hydraulic devices, etc.

(5) Transmission-fluid pumps which provide the hydraulic pressures which engage and disengage the clutches, etc., in (3), move valves within the hydraulic transmission control system, provide the clutch or converter in (1) with fluid, lubricate the transmission, and so on.

Some of these points are discussed in succeeding sections of this chapter, and others in chapters pertaining to control systems, planetary gear systems, or particular types of transmissions.

4. Torque and Speed Characteristics

Certain of the characteristics of a transmission and clutch such as those shown in Fig. 3 differ from the corresponding characteristics of most if not all automatic transmissions. These differences in some instances cause marked differences in the operation of the automobiles in which the transmissions are used.

The clutch shown in Fig. 3 provides no torque multiplication; that is, its TF is 1.0 in power-on or power-off operation when the clutch is engaged. If the clutch is "slipped" while the engine speed is considerably higher than the corresponding automobile speed, the engine may be capable of developing more torque than it would if the clutch were not slipped. Some of this added torque can be transmitted through the slipping clutch, but the clutch does not multiply it; the clutch simply transmits torque which is provided by the engine, or by the driveshaft in power-off operation.

The fluid units (fluid clutch or torque converter) used to serve the functions of a mechanical clutch in automatic transmissions may have power-on torque factors of 1.0, or varying from 1.0 to considerably more than 1.0 depending on the type of unit used and on the conditions under which it is used. Further, the TF in power-on and power-off operations may not be the same even though other conditions are the same. These units are discussed in succeeding sections.

The transmission in Fig. 3 provides a certain TF when in any one of the power-on conditions, that is, Low- or Reverse-gear, etc., power-on condition. When in a power-off condition, the TF is the reciprocal of its value in the corresponding power-on condition.

In automatic transmissions, the power-on TF of the transmission gear system in a particular gear (1st gear, for example) may be 2.5 and the power-off TF in the same gear may be 0.0. In another gear of the same transmission (Low gear, for example), the TF of the gear system in power-on operation may be 2.5, and it may be the reciprocal of this TF in power-off operation.

The TF of a complete automatic transmission, including the fluid unit and gear system, is the product of the torque factors of the two parts, or

$$TF_{\text{trans}} = (TF_{\text{fluid unit}})(TF_{\text{gear syst}})$$

which may have a value equal to or greater than the TF of the gear system in power-on operation. Also, it may be greater than the TF of the gear system in power-on operation and equal to that TF in power-off operation.

The TF of the manual transmission and clutch, likewise, is the product of the torque factors of the two elements, or

$$TF_{\text{trans and clutch}} = (TF_{\text{clutch}})(TF_{\text{gear syst}})$$

which always is the same as that of the gear system if the clutch is engaged and zero if the clutch is disengaged.

The speed factor of the manual transmission in Fig. 3 is a specific value when in any one of its gear positions (Low, etc.) and operating with power on, and it is the reciprocal of this value when operating with power off. The speed factor of the clutch is 1.0 when the clutch is engaged, 0.0 when disengaged, and some value between 0.0 and 1.0 when partially engaged.

Since the torque factor of the gear system of an automatic transmission may be 0.0 in power-off operation in certain of its gear positions, the SF is 0.0 in these cases. (This SF pertains to the speed of the "new" output shaft which is caused by the rotation of the "new" input shaft, and disregards the speed of the "new" output shaft which results from the power provided to it by the engine.) The fluid unit used in automatic transmissions in lieu of the mechanical clutch of Fig. 3 may be considered to be engaged at all times, but the engagement is through a fluid rather than a mechanical connection. Its SF may have any value from 0.0 to approximately 1.0 in power-on or power-off operation, depending on the relationship between the torque required by the load connected to the fluid unit output shaft and the difference in speeds of the parts of the unit that is required to produce this torque.

As in the case of torque factors,

$$SF_{\text{auto trans}} = (SF_{\text{fluid unit}})\,(SF_{\text{gear syst}})$$

which may have any value from 0.0 to a value almost equal to the SF of the gear system.

Likewise,

$$SF_{\text{man trans and clutch}} = (SF_{\text{clutch}})\,(SF_{\text{gear syst}})$$

which is the same as that of the gear system when the clutch is engaged, zero when the clutch is disengaged, and between zero and $SF_{\text{gear syst}}$ when the clutch is partially engaged.

Chapters 11–20 include the torque and speed factors of the gear systems of various types of transmissions and of the complete transmissions including the fluid units. In these chapters the gear designations include terms such as 1st, 2nd, 3rd, Low, High, First, Second, and Neutral. These indicate the condition of the gear system that exists or may exist when the transmission control device is in the Drive, etc., position.

The designations 1st gear, Low gear, and First gear usually are synonymous terms in manually controlled transmissions and indicate the same transmission speed and torque factors. The values of these factors in power-off operations usually are the reciprocals of those in power-on operations. As stated before, this reciprocal relationship as it pertains to the gear portion of automatic transmissions may or may not exist, depending on the position of the transmission control device and the design of the transmission. Chapter 12 states that the gear portion of the Torque-Flite transmission has a power-on torque factor of 2.45 when the control device is in the First position, and may have this same factor when in the Drive or Second position. The power-off torque factor in the First position is 0.408, which is equal to $\frac{1}{2.45}$. The power-off torque factor in the other two positions and in the gear corresponding with First gear is 0.0. For this reason (and other reasons stated below), the designation 1st gear is used when the transmission control device is in the Drive or Second position, and First gear is used when the control is in the First position.

23

The second reason for use of different terms is that the conditions of some of the bands and clutches in the transmission (Fig. 24) are not the same when in 1st gear and First gear. It is this difference which causes the reciprocal TF and SF relationships to exist or not exist.

The third reason is that the transmission will not remain in 1st gear in power-off operation when the automobile speed is above a few miles per hour or in power-on operation when the automobile speed is above several miles per hour, whereas it will not shift out of First gear in power-off or power-on operations. A part of this reason applies also to the use of the terms 2nd gear and Second gear.

Some of the above reasons cause the terms Low gear and 1st gear to be used in other transmissions. In two-speed transmissions (Chapter 13 and others), the terms Low, 1st, and High gears are used. High gear corresponds to 3rd gear in three-speed transmissions.

The approximate ranges of some of the power-off speed factors of the complete transmission in Chapters 11–20 are 2.45–0.0, or some other value to 0.0. The 0.0 and near 0.0 values pertain to ignition-off operation or to operations in which the engine is being started by pushing the automobile or other similar means. Otherwise, the engine would drive the "new" output shaft of the transmission at low or zero automobile speeds in any of the transmission control positions.

When the transmission control device of an automatic transmission is in the Park position, the output shaft of the transmission is locked to the transmission case and cannot rotate. Therefore, no torque can be transmitted through the transmission and the TF and the SF of the transmission are both 0.0. Also, the conditions of the gear portion of the transmission are the same as when in the Neutral position so the TF of this part of the transmission is (substantially) zero and the SF is zero.

5. Fluid Clutches

a. GENERAL DESCRIPTION. Figure 4 is an illustration of a fluid clutch. The clutch consists of an impeller and a runner (sometimes given other names) enclosed in a housing. It (usually) is full or almost full of transmission fluid, or clutch fluid if the clutch is not integrated with the transmission in this respect. The fluid may remain in the housing at all times or it may be removed from and returned to it on a continuing basis in order to remove excess thermal energy and reduce the high temperatures that result from friction—primarily, friction between "drops" of the fluid. This type of friction is known as viscous friction.

The input shaft may be connected to the engine, in which case the input shaft usually is the engine crankshaft. In some transmissions, the clutch is placed elsewhere in the power-flow path so the input shaft is not the engine crankshaft, nor is it connected directly to the crankshaft. The output shaft usually, if not always, is connected directly or indirectly to a gear of the transmission.

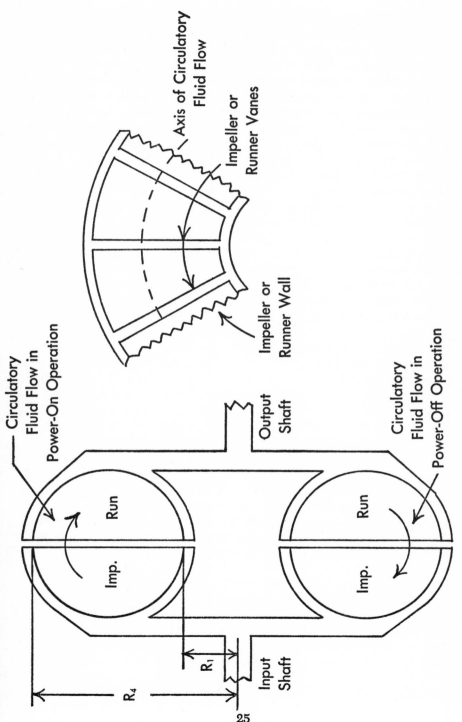

Axis of Circulatory Fluid Flow

Impeller or Runner Vanes

Impeller or Runner Wall

Circulatory Fluid Flow in Power-On Operation

Run

Imp.

Output Shaft

Circulatory Fluid Flow in Power-Off Operation

Run

Imp.

R_1

R_4

Input Shaft

25

Fig. 4. Fluid clutch. Clutch housing is not shown.

In the following it is assumed that the input shaft of the clutch is the engine crankshaft. When the engine is running and the clutch runner has a rotational speed which is less than that of the engine and impeller, the impeller operates as a centrifugal pump and pumps the fluid from the points which are near its axis of rotation to points which are farther from that axis. At these points the fluid leaves the impeller and moves through the runner and back to its starting points in the impeller. From here the fluid repeats its cycle of movement through the impeller and runner over and over again as long as the impeller speed is greater than that of the runner.

When the runner rotates it also acts in the manner of a centrifugal pump, but when the impeller speed is greater than that of the runner the pumping action of the impeller exceeds that of the runner and the fluid flows as described above. When the speeds of the two elements are equal, no fluid flow occurs between these elements. When the runner speed is greater than that of the impeller the runner acts as an impeller and the impeller acts as a runner, and the fluid flows between the elements in the manner described but in the opposite direction. The runner speed is greater than that of the impeller when the conditions are such that the automobile tends to drive the engine, that is, in power-off operations.

The rotation of the impeller, when its speed is greater than that of the runner, causes the fluid in the impeller to rotate around the axis of rotation of the clutch at the same rotational speed as that of the impeller. When the fluid leaves the impeller and strikes the vanes of the runner it produces a force on those vanes. This force produces a torque on the output shaft of the runner which is applied to a gear of the transmission and to the driven wheels of the automobile if the transmission is "in gear." If the runner speed is greater than that of the impeller, the opposite occurs and the torque on the impeller is applied to the engine crankshaft in the direction which tends to increase the speed of the engine. The direction of rotation of the fluid around the axis of the clutch remains the same as when the impeller speed is the greater speed but, as stated above, the direction of flow in planes which include the axis of rotation of the clutch is in the opposite direction.

b. Fluid Flow. Assume the following:

(1) The space within the runner and the impeller is completely full of transmission fluid, and the fluid is incompressible.

(2) The edges of the walls of the impeller and the runner are so close to one another that the amount of fluid that enters and leaves the enclosure formed by these elements within a period of a few seconds is negligible.

(3) The speed of rotation of the impeller, S_{imp}, is greater than the

speed of the runner, S_{run}. (Throughout the following, rotational speeds are in revolutions per minute, rotational velocities are in radians per second (Chapter 1, § 3), and linear or tangential velocities are in feet per second.)

(4) The input shaft of the impeller is the engine crankshaft.

(5) The impeller and the runner are symmetrical insofar as the walls and the vanes are concerned.

Consider the volume of fluid between two adjacent impeller vanes. The distances between these vanes at the outer and inner extremities of the vanes, in feet, are $2\pi R_4/N$ and $2\pi R_1/N$, where R_1 and R_4 are the radii indicated in Fig. 4, and N is the number of impeller vanes in the clutch, assuming the thickness of each vane to be negligible in comparison with the spacing between the vanes. The fluid flow in the outer portion of this volume is in the direction toward the runner, while that in the inner portion is away from the runner. Along a circular line in the vicinity of the centers of the circles formed by the walls of the impeller and the runner, the fluid does not move toward or away from the runner. This line is the axis of rotation of the fluid as it circulates through the impeller and the runner.

Any very small volume of the fluid, such as the volume occupied by a drop of the fluid, has a velocity component due to the above flow between the impeller and the runner and another velocity component due to the rotation of the impeller or the runner around its axis. These velocity components may be expressed as rotational and linear components, and are:

W_r, The rotational velocity around the axis of the clutch, or the radial-plane velocity, in radians per second,

V_r, the linear, or tangential, velocity corresponding with W_r, in feet per second,

W_c, the rotational velocity in planes which include the axis of the clutch which is the velocity of the small volume between the vanes of the clutch, or circulatory-plane velocity, in radians per second,

V_c, the linear, or tangential, velocity corresponding with W_c, in feet per second.

The rotational velocity components of a small volume of fluid have axes that are 90 degrees from one another. Likewise, the linear velocity components at any instant have directions which are 90 degrees from one another. The linear velocity of the volume is the vectorial sum of the two linear velocity components.

The numerical values of the linear velocities pertaining to all of the drops of fluid differ from one another by considerable amounts in many

27

instances. Consider very thin layers of fluid, with thicknesses comparable with the diameter of a molecule of fluid, which are in contact with the walls and the vanes of the clutch. The velocity of these layers with respect to the walls and the vanes is zero because of the strong forces of adhesion which act between the fluid and the metal surfaces. The V_c velocity component of the fluid at the axis of circulatory movement also is zero. The circulatory velocity components of other parts of the fluid increase with distance from these places and are greatest in the vicinity of the midpoints between the vanes and the midpoints between the axis of circulatory movement and the impeller-runner wall.

The fluid in other very thin layers which are in contact with or very close to the above layers usually moves in a layer fashion. Likewise, the remainder of the fluid may move in thin layers. However, in many cases as indicated below a large part of the fluid moves in a turbulent manner. When this occurs, the only part of the fluid which moves in layers is that part which is near the zero-velocity "drops" of fluid.

While the rotational and linear circulatory-plane velocity components of the various parts of the fluid differ greatly, there is an average of the V_c velocities of all of the drops which can be considered to be the V_c velocity of all of the fluid. Likewise, there is a W_c velocity corresponding with this V_c velocity that can be considered to be the W_c velocity of all of the fluid.

Any combination of the pumping forces that are produced by the rotations of the impeller and the runner would cause particular velocities, W_c and V_c, of all of the fluid if viscous friction did not exist. Viscous friction has a property which differs from that of friction between solids in that the force of viscous friction is directly proportional to the difference in the relative velocities of the surfaces of the layer of fluid between which the movement occurs in those instances in which the fluid flows in layers. Thus,

$$F_t = u\,A\,V_s/h$$

where F_t is the force of friction, u is the viscosity of the fluid, A is the area of each of the surfaces of the layer (or the average of the areas), V_s is the difference between the relative velocities of the surfaces, and h is the thickness of the layer.

When the velocity and the viscosity of the fluid and other factors have certain relationships, most of the fluid does not flow in layers but flows in a turbulent manner. The relationship between F_t and the factors which affect it is complex when turbulent flow exists. The flow in fluid clutches is turbulent when the difference between the speeds of the impeller and the runner is not small.

The relationship of the resultant of the combination of forces which

28

produce a fluid flow, per square foot of cross-sectional area of the fluid, and V_c in turbulent or laminar flow (Chapter 1, § 3) is

$$V_c = K(2F_c/(Am))^{1/2}$$

where V_c is in feet per second, m is the mass of the fluid per cubic foot, F_c/A is the force per square foot, and K is a factor which is determined by the shape of the fluid-flow passageway, the viscosity of the fluid, and other items as indicated in the above discussion of F_f. V_c is the average of the circulatory-plane velocities of all of the drops of fluid.

c. TORQUE FACTOR. The torque that is produced on the runner by the fluid is caused in a direct fashion as the result of the V_r or W_r velocity of the fluid. It is caused in an indirect manner by the V_c or W_c velocity.

Chapter 1, § 3, states that the torque required to cause an angular acceleration of an object is

$$Tq_{in} = M/t\, R_i^2\,(W_a - W_b)$$

where M is the mass of the object which is its weight in pounds divided by 32.2, R_i is the radius of gyration or the radius of the center of inertia from the axis of rotation in feet, t is the time in seconds during which the acceleration occurs, W_a and W_b are the angular velocities of the object in radians per second at the end and beginning of the time t, and Tq_{in} is in lb-ft.

When a small volume, or drop, of the fluid in a clutch is considered, the value of R_i changes continuously as the drop rotates with a velocity of W_c, where this velocity is the velocity of the drop. Therefore, it is more convenient to use another torque equation that pertains to cases such as this. It is

$$Tq_{in} = M_d/t\,(W_{imp}\,R_{ii}^2 - W_{run}\,R_{ir}^2)$$

where M_d is the mass of the drop, W_{imp} and W_{run} are the rotational velocities of the impeller and the runner, R_{ii} is the radius of the drop when it leaves the impeller, and R_{ir} is the radius when it leaves the runner.

When all of the fluid is considered, the equation is

$$Tq_{in} = M/t\,(W_{imp}\,R_{i3}^2 - W_{run}\,R_{i2}^2)$$

where M/t is the mass of fluid that passes through the impeller per second, R_{i3} is the radius of gyration of all of the drops of fluid when leaving the impeller, and R_{i2} is this radius when the drops are entering the impeller. These radii are approximately

$$R_{i3} = ([R_4 + R_1]/2 + R_4)/2 = (3R_4 + R_1)/4, \text{ and}$$
$$R_{i2} = ([R_4 + R_1]/2 + R_1)/2 = (R_4 + 3R_1)/4$$

The torque equation also is that of the torque which is produced on the runner by the fluid since the mass of fluid that flows through it per second is the same as that flowing through the impeller per second and since the radii are the same except that R_{13} is the radius of entering and R_{12} is the radius of exit. Then, assuming no torque loss in the clutch,

$$Tq_{out} = Tq_{in}, \text{ and}$$
$$TF = Tq_{out}/Tq_{in} = 1.0$$

The energy losses within a fluid clutch which affect the output torque are small, so

$$TF = \text{slightly less than 1.0 except as stated below}$$

The output torque is zero when the impeller and the runner velocities have nearly-equal values since the pumping forces produced by these elements are nearly equal and M/t becomes very small. Under these conditions, all of the input torque is used in balancing the effects of the energy losses, so Tq_{out} and TF are 0.0.

d. Torque Capability. The output torque of the clutch, from the above, is

$$Tq_{out} = TF \, Tq_{in}$$

In normal operations, TF has values very near its theoretical value, which is 1.0. The radii of gyration are determined by the dimensions of the clutch. Therefore, when W_{imp} and W_{run} are maintained constant in a particular clutch, the value of Tq_{out} is determined by the value of M/t.

The value of M/t is a function of the average of the V_c velocities of all of the drops of fluid and of the mass of the fluid per cubic foot. It also is a function of the amount of fluid that is in the clutch in those instances in which the impeller and the runner are not full of fluid.

The circulatory velocities are determined by the relationships between the forces which increase those velocities and those which decrease them. These forces include (1) the centrifugal forces that result from the velocity, W, of the impeller, (2) the centrifugal forces that result from the velocity, W, of the runner, (3) the forces of viscous friction, and the forces which result from such items as the curved flow path and the nonuniform cross-sectional areas of the fluid as the fluid moves along the path.

The first of these tends to produce circulatory velocities in one direction, the second tends to produce velocities in the opposite direction, and the third reduces the velocities to values which are less than those that would result if the forces of friction and the other forces did not exist.

When the velocity of rotation of the runner is zero, the forces in (2) are zero. If the velocity of the runner equals that of the impeller, the

30

forces in (1) and (2) are equal and the circulatory velocities are zero. This condition cannot exist except in a transient manner unless the load provides power to the runner since friction forces always are present and the impeller velocity must be greater than that of the runner (or vice versa) in order to produce a torque to balance the effect of these forces.

The centrifugal force of a drop of fluid in the impeller if its V_c velocity is zero (Chapter 1, § 3) is

$$F_{c\,imp} = M_d\,V_r^2/R$$

where V_r is the radial-plane velocity and R is the radius of the drop from the axis of rotation of the clutch. This force is due to the constant acceleration of the drop toward the axis of rotation, which is the constant change in the direction of the V_r velocity of the drop. When the drop follows a (nearly) circular path due to its W_c velocity, its V_c velocity changes direction constantly. Therefore, there is an acceleration of the drop toward the axis of circulatory movement as the result of the W_c, or V_c, velocity. As stated before, the V_c velocities vary considerably from drop to drop. Therefore, the centrifugal force which results from the V_c velocity varies greatly from drop to drop in many instances.

The total acceleration of any drop is the sum of its acceleration components. Therefore, the value of the force acting on the drop is not that in the equation above if the V_c velocity is not zero. The same factors affect the force acting on a drop when the drop is in the runner. Further, the values of the forces which oppose the resultant of the impeller and the runner pumping forces, caused by viscous friction and other items, have complex relationships with V_c and many other factors, as discussed previously. For these reasons, the values of V_c and W_c, and the resultant values of M/t and the torques that may be expected in newly designed (but not constructed) clutches probably can be determined best by the use of data pertaining to existing clutches which resemble the new clutch.

The equation for M/t in a centrifugal pump (or impeller of a clutch when the runner velocity is zero) may be expressed as

$$M/t = K_p\,D_{imp}^{\,3}\,W_{imp}$$

where K_p is a factor which pertains to the particular design of the pump, D_{imp} is the maximum diameter of the vanes of the pump impeller in feet, and W_{imp} is the velocity of the impeller in radians per second. Then, using this relationship in the clutch torque equation when W_{run} is zero,

$$Tq_{in} = K_{clutch}\,D_{imp}^{\,3}\,W_{imp}^{\,2}\,R_{i3}^{\,2}$$

or the same equation with W_{run} substituted for W_{imp} when power-off conditions exist.

31

Similar clutches may be defined as those which use the same type of fluid, are similar in construction, and in which the ratio of any two dimensions (such as R_4/R_1) in one clutch is equal to that in the others. Then

R_{i2} and R_{i3} may be expressed as $R_{i2} = C_1 D_{imp}$, and
$$R_{i3} = C_2 D_{imp}$$

where the values of C_1 and C_2 are the same in all of the similar clutches. When the runner speed is zero use of the value of R_{i3} in the torque equation above results in

$$Tq_{out} = TF\, Tq_{in} = TF\, C\, S_{imp}^2\, D^5$$

where S_{imp} is the speed of the impeller in rpm, D is the maximum diameter of the impeller in feet, and TF and C are (substantially) the same in all of the similar clutches. The value of TF is substantially 1.0. A representative value of C is 0.0002 when the ratio of R_4/R_1 is approximately 3/1, a ratio which frequently is used in clutches.

The transmission shown in Fig. 33 has two fluid clutches. One of these has no fluid in it when the transmission is operating in two of its four forward-speed gears, and is full of fluid when in the other two gears. When being filled or emptied, the torque of the clutch varies in accordance with the amount of fluid it contains. Thus, C in the torque equation is changed by the gear-shifting control system.

The quantity, $(S_{imp} - S_{run})/S_{imp}$, where S_{run} is the speed of the runner in rpm, is known as the slip of the clutch. Frequently, it is expressed as a percentage, in which case it is

$$\text{Slip} = 100\% \ (S_{imp} - S_{run})/S_{imp} = 100\% \ (1 - S_{run}/S_{imp})$$
$$= 100\% \ (1 - SF)$$

where SF is the speed factor.

When the slip is zero, $S_{imp} = S_{run}$ and Tq_{out} is zero. When the slip has some very small value, the output torque and the TF are zero since all of the input torque is used to balance the torque required by the energy losses in the clutch. The relationship between the values of Tq_{in} and slip when S_{imp} is constant is similar to that shown in Fig. 5. As indicated before, the values of the fluid-flow factor, K, vary with the designs of the clutches and are difficult to compute. Therefore, the values of Tq_{in} vs. slip in percentages and in numerical quantities are determined by tests of the clutch, or they may be predicted during the design period of a new clutch by the use of data pertaining to clutches which resemble the new clutch.

e. POWER RELATIONSHIPS AND SPEED FACTOR. The power provided to the input shaft of the clutch (Chapter 1, § 3) is

$$P_{in} = Tq_{in}\, S_{in}/5{,}252$$

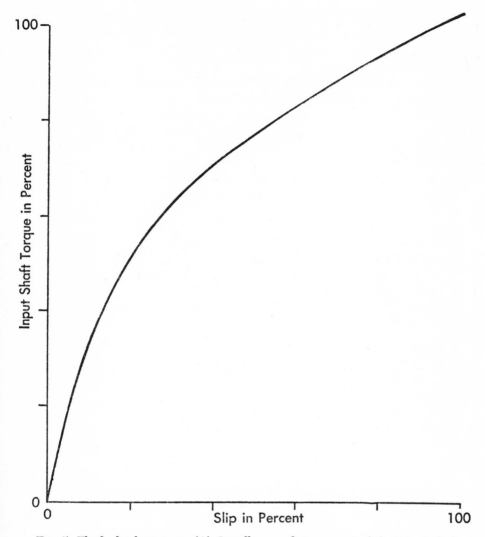

FIG. 5. Fluid clutch torques. (1) Impeller speed is constant. (2) Output shaft torque is (TF) (input shaft torque). (3) TF is substantially 1.0 except when slip is very small, when it is 0.0.

where P_{in} is in horsepower, Tq_{in} is in lb-ft, and S_{in} is the speed of the engine in rpm, assuming that the engine crankshaft is connected directly to the impeller.

The horsepower output from the clutch to its load is

$$P_{out} = Tq_{out}\, S_{out}/5{,}252$$

where S_{out} is the speed of the runner.

Since the torque factor is substantially 1.0 except when the slip is very small,

$$P_{out}/P_{in} = S_{out}/S_{in}$$

except when S_{out}/S_{in} is almost 1.0. This power ratio is zero when the output shaft is stalled, slightly less than 1.0 when the slip is a few per cent, and zero when the slip is such that the output torque is zero. Thus, when the output shaft is stalled, all of the output power of the engine is transformed into thermal energy in the clutch and must be dissipated in the form of heat. When the slip is small, the thermal energy produced is much less than when the slip is great for any given value of S_{in}.

As indicated above, the speed factor of a fluid clutch,

$$SF_{clutch} = S_{out}/S_{in}$$

varies from 0.0 to nearly 1.0. The value of the SF at any instant depends on the relationship between the torque required by the load and the slip required to produce that torque. The SF cannot equal 1.0 (except in a transient manner) even with no load since the torque required because of energy losses makes a small slip necessary. However, the SF may be considered to be 1.0 under this condition with a very small error.

f. Power-off Operation. Power-off operations are those in which the runner shaft is the "new" input shaft and the impeller shaft is the "new" output shaft. The torque and power flows, then, are in a direction opposite to the usual direction.

As indicated before, the impeller and the runner are very similar. Therefore, the torque factors and the speed factors, the amounts of torque transmitted through the clutch, etc., are the same or substantially the same in power-on and power-off operations under corresponding load, speed, etc., conditions. The principal differences insofar as the clutch is concerned are that the runner serves as an impeller and vice versa, and the direction of the fluid flow in the circulatory planes is not the same in the two cases.

g. Combined Clutch and Automobile Characteristics. Figure 6 illustrates the speed-vs.-torque characteristics of an automobile and its engine, and the torque characteristics of a fluid clutch which couples the engine and the gears of the transmission. Assume that the transmission has two forward-speed gear ratios and that the control lever is in the Low position which causes the TF and the SF of the gear system to be 2.0 and 0.5, respectively, in power-on operation. Assume, further, that the automobile is to be started on a steep up-grade road with the greatest possible acceleration to a high speed, and that the tires can provide more traction than is needed.

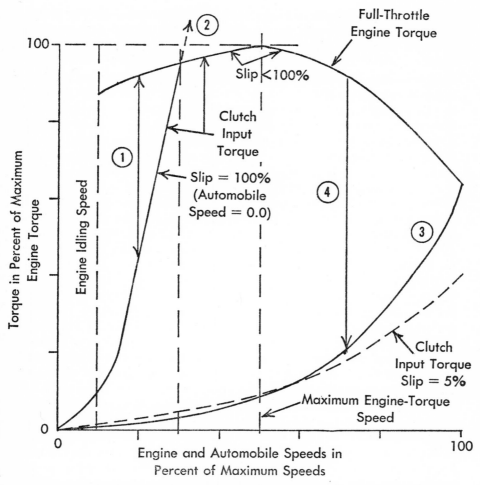

Fig. 6. Fluid clutch and automobile torques. (1) Torque available for engine speed acceleration. (2) Clutch input torque if not limited by engine torque. (3) Clutch input torque required in high gear on level road with constant automobile speed. (4) Torque available for automobile acceleration in high gear.

Note: Clutch Output Torque = (TF) (Input Torque)
= Approximately Input Torque

Under the above conditions, the throttle is moved to the fully opened position. Before the throttle was opened from the idling position, the maximum output torque the engine could produce was substantially equal to that the runner was capable of transmitting to the gear part of the transmission at a runner speed of zero with the engine idling. The maximum opening of the throttle causes the engine to have a torque capability at its idling speed which is many times greater than the

35

capability before the throttle was opened. This torque cannot be transferred to the impeller and the runner at the idling speed since the difference between the speeds of the clutch elements is small even though the slip is 100 per cent. Therefore, the engine speed increases very rapidly to approximately 60 per cent of its maximum-torque speed. The increase in speed causes the difference in speeds of the clutch elements to increase and the runner to receive and transmit much greater torque to the gears. This greater torque accelerates the automobile and the acceleration continues until the torque required from the engine by the automobile at the speed attained is equal to the maximum torque capability of the engine at its speed. If the up-grade road is sufficiently steep, the engine will not rotate faster than its maximum-torque speed. Likewise, the automobile will not accelerate beyond the speed corresponding with this engine speed, the Low-gear speed factor of 0.5, and the clutch slip required to transmit the torque required from the clutch by the automobile at its speed. Thus, the fluid clutch torque characteristics are such that the engine rotates at a speed corresponding with a large engine-torque capability when the load requires that torque. Similar results can be obtained with a mechanical clutch if the clutch is "slipped" skillfully, but the heat and the abrasion produced by the slipping for even a short period of time probably would damage the clutch if the slipping were done under heavy load conditions.

The general conditions are the same as in the above in operations on level roads with moderate acceleration to the cruising speed. A partial opening of the throttle causes the engine to rotate considerably faster than the runner when the automobile is starting. Then the difference between the speeds of the impeller and the runner decreases until it reaches the minimum value it can have and provide the torque required at the maximum desired speed in Low gear. If the transmission control lever is placed in the Drive position at this speed, the automobile speed will increase, assuming that the engine speed is maintained constant since the SF of the gear portion of the transmission in High gear is 1.0. The acceleration of the automobile requires an increase in the output torque of the clutch. The shift of gears changed the TF of the gears from 2.0 to 1.0, so the torque required from the clutch is still greater because of the change from Low to High gear. The slip of the clutch increases and this permits the clutch to receive and transmit a greater torque. As the automobile approaches the desired cruising speed, the torque required for acceleration decreases and the slip decreases until it becomes a few per cent (perhaps 3 to 5 per cent) at the cruising speed.

Conditions similar to the above exist in power-off operations. If the throttle is closed when the automobile is moving at a high speed on a level road, the engine speed decreases until the torque transmitted to the impeller from the runner prevents a further decrease in engine speed. The torque transmitted to the engine would decrease rapidly as the automobile speed decreases if the engine speed remained constant, so the engine speed decreases until a balance between the torque required by the engine and the torque provided by the clutch is reestablished. This process of automobile and engine deceleration is a continuous one, rather than the step-by-step process that may be indicated by the preceding sentence. When the automobile speed is low (say, 20 mph), the slip starts to decrease because the engine is rotating at a speed which is near its idling speed and the torque transmitted to the engine decreases. The torque continues to decrease until it becomes 0.0 when the runner speed and the impeller (and engine) speed become very nearly equal. If the ignition had been turned off when the throttle was closed, the torque transmitted to the engine would have been greater throughout the above deceleration and it would have become zero when the runner speed and the automobile speed became very nearly zero, assuming that the transmission did not shift to Neutral gear. Chapter 1, § 3 discusses the power and torque required to drive an engine in power-off operations.

Since the power-off torque factor of a fluid clutch is substantially 1.0 (except as indicated in the preceding paragraph) and the speed factor is less than 1.0, the power transferred from the automobile to the engine is less than that the automobile would transmit if a mechanical clutch were used and all other conditions remained unchanged. Therefore there is less "braking power" due to the engine friction and less deceleration of the automobile than there would be if a mechanical clutch were used. The extent to which the "braking power" is less is determined by the speed factor of the clutch. When the torque required to drive the engine at a speed near the speed of the driveshaft (when the SF of the gears is 1.0) is greater than the clutch can transmit with a small slip, the engine decelerates until the torque required to rotate it is equal to the torque-transmitting capability of the clutch. The slip required for this balance of torques causes the power transmitted from the automobile to the engine to be reduced a corresponding amount and the "braking power" of the engine is reduced accordingly.

h. DESIGN FEATURES. Some fluid clutches have hollow centers in the impellers and the runners similar to the one in the torque converter shown in Fig. 7. The term, hollow center, is intended to indicate that the center is hollow insofar as the vanes are concerned, rather than indicating that the center contains no fluid. This clutch design feature reduces the turbulence of the fluid that occurs near the centers of the circles formed by the walls of the impeller and the runner in the clutch shown in Fig. 4. This turbulence is the result of the fluid leaving the impeller near the centers of the circles, entering the runner and almost

37

immediately leaving it and returning to the impeller, and crossing the gap between these elements twice during each cycle of its movement. Some of the fluid probably "splashes" in various directions in and near the gap, striking the runner and the impeller vanes and forming small eddy currents in the process of splashing, thereby causing power losses. These losses may be a large percentage of the power transferred to the runner by the fluid which circulates in this part of the clutch. Therefore, the power efficiency of the part of the clutch near the centers of the circles formed by the impeller and the runner walls may be low. The hollow center prevents fluid flow between the impeller and the runner at and near these centers and thereby improves the power efficiency of the clutch.

The numbers of vanes in the impeller and the runner usually differ by about 10 per cent of the number in one of these. The reason for this is indicated in the following. If the numbers are the same, the vanes of the two elements provide unobstructed passageways for the circulatory flow of fluid when they are aligned with one another. A fraction of a second later, the vanes are not aligned and the fluid strikes the edges of the vanes as it goes from the impeller to the runner, and vice versa. The changes from unobstructed to obstructed to unobstructed paths tend to produce variations in the fluid velocity. While this may not cause undue mechanical stresses in the clutch or undue torque variations, it may produce an objectionable noise, or hum. When the numbers of vanes differ by an appropriate amount, the total of the effective cross-sectional areas of the fluid paths provided by the sets of vanes is substantially constant. This reduces the objectionable effects of equal numbers of vanes.

i. SUMMARY. The torque factor of a fluid clutch in power-on or power-off operations is slightly less than 1.0 except when the slip is very small.

The speed factor in power-on or power-off operation varies from 0.0 to slightly less than 1.0, depending on the value of the torque that is required by the load and the slip required to produce that torque. The torque requirement placed on the clutch by an automobile is a function of the automobile speed and acceleration, the TF of the gear portion of the transmission, and of other factors such as road conditions.

The output torque of any given clutch in power-on operations is proportional to the square of the speed of the impeller when the runner is stalled. It becomes zero, and the torque factor becomes zero, when the runner speed is slightly less than the impeller speed. The power-on or power-off torque-vs.-slip relationship is similar to that illustrated in Fig. 5.

The maximum output torques of clutches which are the same except that all of the dimensions of one are a multiple of the corresponding dimensions of another may be expressed as

$$Tq_{out} = TF \, C \, S_{imp}^2 \, D^5$$

where TF and C are (substantially) the same for all of the similar clutches, S_{imp} is the speed of the impeller in rpm, and D is the diameter in feet of the impeller of the clutch being considered. The value of C is determined by the design of the similar clutches, but a representative value is 0.0002. The maximum output torque in power-off operations is the same as that in power-on operations when S_{run} is substituted for S_{imp}.

The ratio of P_{out}/P_{in}, which is the power efficiency, is substantially equal to the ratio, S_{out}/S_{in}, which is the speed factor, except when the speed ratio is almost 1.0. The power efficiency is zero when the runner (or impeller) is stalled, and zero when the slip is such that the output torque is zero. It is a maximum when the slip is such that

$$P_{out}/P_{in} = P_{out}/(P_{out} + \text{power losses in the clutch})$$

is a maximum. This occurs when the slip is a few per cent.

6. Fluid Torque Converters

a. GENERAL DESCRIPTION. Figure 7 is used to illustrate the principles of a three-element torque converter. In some respects, a torque converter resembles a fluid clutch; in others, it differs from a clutch. The differences cause the torque characteristics of the two fluid units to be considerably unlike.

The impeller shown in Fig. 7 differs from that in Fig. 4 in that it has a hollow center, but, as stated in § 5, some fluid clutches also include this feature. Therefore, the impellers of torque converters and fluid clutches are similar to one another, although the vanes of converter impellers usually have some curvature.

The runner in a fluid clutch corresponds with the turbine (sometimes given other names) in the torque converter. They differ in that the surfaces of the vanes of a runner are straight, while those of a turbine have considerable curvature.

The stator in a converter has no corresponding element in the clutch. It includes curved vanes between which the fluid flows in going from the turbine to the impeller in power-on operation, or in the opposite direction in power-off operation. The stator has a shaft which can rotate freely in the forward direction, which is the direction of rotation of the input shaft. When the stator tends to rotate in the reverse direction, the rotation is prevented by an overrunning clutch, sometimes called a one-way clutch. This clutch operates in a manner similar to that of a ratchet

39

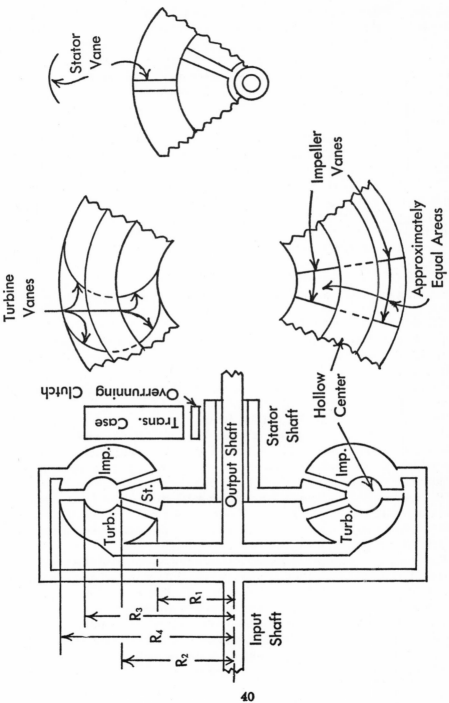

FIG. 7. Three-element torque converter. Converter housing is not shown.

40

wheel and its stop, but the clutch has no teeth in the usual sense of that word. The transmission case serves as the stationary object to which the ratchet stop would be attached if a ratchet were used.

Three-element converters are used extensively in automatic transmissions but converters with more elements are used also. One such converter is shown in Fig. 30.

Torque converters usually are placed in the power train between the engine and the gear portion of the transmission. (In Fig. 30 a gear set is a part of the converter, both structurally and functionally.) The input shaft of the converter, then, is the engine crankshaft, and the output shaft is the input shaft of the gear portion of the transmission.

The stator of the torque converter in Fig. 26 is used as a turbine in Reverse-gear operation.

A large part of the discussion in § 5 is applicable to torque converters since the principles involved in the operation of fluid clutches and converters are the same or similar.

b. Fluid Flow. For the purposes of the following discussion, assume that the input shaft of the converter shown in Fig. 7 is connected to a power source that is controlled in a manner such that the output power of the source remains constant as the speed changes through a wide range. This power source is used to avoid the variation in power capability of an automobile engine with speed which introduces variables that would confuse the discussion. The effects of the use of an automobile engine are discussed later. Also assume that the output shaft of the converter is stalled—that is, its speed is and remains zero unless indicated otherwise.

For the purposes of analysis, let it be assumed that a very small volume of transmission fluid is about to enter the impeller from the stator at points along the wall of the impeller which are nearest to the axis of rotation. Further assume that no other fluid is in the converter. Let the linear velocity of the small volume of fluid be V, and the components of this velocity be V_{r1} and V_{c1}, where V_r is the velocity in the planes of rotation of the converter, V_c is the velocity in planes which include the axis of rotation, and the numeral 1 indicates the velocities when the radius in feet of the volume from the axis of rotation is its minimum value, R_1, when the volume is in the impeller or the turbine.

The impeller transmits energy from the power source to the small volume of fluid as it moves from its points of entry into the impeller to its points of exit from the impeller. Assume that the latter points are at the greatest possible radius from the axis of rotation. The first component of the added kinetic energy, expressed in ft-lb and due to the velocity component, V_r, assuming that the fluid entered the impeller with zero velocity in the radial planes (Chapter 1, § 3), is

41

$$KE_{comp\,1} = M/2\,V_{r4}^2$$

where the subscript 4 indicates the radius, R_4, in feet, M is the mass of the small volume of fluid, and the velocity is in feet per second.

If the velocity of the fluid in the radial planes immediately prior to entering the impeller is not zero, the general expression of the added kinetic energy is

$$KE_{comp\,1} = M/2\,(V_{r4}^2 \pm V_{r1}^2)$$

where V_{r1} is the velocity immediately prior to entry into the impeller at radius R_1. In stalled operating conditions, and in other conditions in which the impeller speed is considerably greater than the turbine speed, the torque applied to the stator by the fluid is in the reverse direction and the overrunning clutch holds the stator stationary. Under these conditions, the shape and the angles of the stator vanes with respect to the flow of fluid are such that the second velocity in the expression always is in the same direction as that of the first, or forward direction, so

$$KE_{comp\,1} = M/2\,(V_{r4}^2 - V_{r1}^2)$$

Figure 8 is intended to serve as an aid in understanding the influence of the converter elements upon the fluid velocities and vice versa. It shows an "in-line torque converter" which is impractical from the standpoint of use as a converter, but the above influences may be easier to understand by the use of Fig. 8 than of Fig. 7. The three sets of impeller vanes in Fig. 8 (only two vanes of each set shown) represent the one set in Fig. 7. Likewise, the two sets of turbine vanes and stator vanes shown

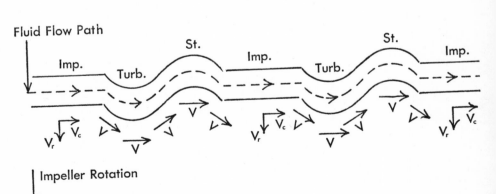

FIG. 8. Simulation of three-element torque converter. (1) All impeller vanes connected to one input shaft and all turbine vanes to one output shaft (behind vanes in illustration). (2) All stator vanes held stationary or permitted to rotate in forward direction. (3) V_r and V_c are radial-plane and circulatory-plane components of fluid velocity, V.

42

represent the single sets in Fig. 7. The velocity vectors, V_r, V_c, and V, correspond with those used in the discussion of the converter in Fig. 7, and indicate the directions of the fluid velocity components and the resultant of those components at various points within the converter.

The second component of the fluid velocity, V_c, has values which are dependent on the relative values of the force which tends to increase the velocity of the fluid in the planes which include the axis of rotation of the converter, and the forces which tend to decrease that velocity. For the purposes of this discussion it is not necessary that the values of this quantity be expressed in a quantitative manner. Therefore, the velocity of the small volume of fluid is expressed as V_c, V_{c4}, etc. Section 5 indicates the relationship between V_c and W of the impeller and the turbine.

c. TORQUE AND TORQUE FACTORS. When the small volume of fluid leaves the impeller, it has linear velocity components of V_{r4} and V_{c4}. Assume that the relative magnitudes of these components are such that the resultant velocity, $V_{out\ imp}$, is in a direction which causes the small volume to enter the turbine along a path tangential to the concave surface of a vane of the turbine at the points of entry. Assume, further, that the successive directions of the resultant velocity, V, pertaining to very short distances of travel of the fluid form the arc of a circle as the small volume travels through the turbine. Then this velocity has an acceleration toward the center of a circle of which the turbine surface forms an arc. The force exerted by the fluid on the turbine at any instant during its movement through the turbine (Chapter 1, § 3) is

$$F = M V_{turb}^2/R_v$$

where F is in pounds, V_{turb} is the velocity in feet per second of the volume of fluid at the instant being considered, and R_v is the radius of curvature of the V velocity in feet. Since the turbine is stalled, the only forces tending to change the magnitude of the velocity of the fluid are the forces of viscous friction which act within the small volume and the forces which act between the surface of the turbine and the surface of the fluid. For purposes of discussion, let it be assumed that these forces are small. Then, $V_{out\ turb}$ and $V_{in\ turb}$ are nearly equal in magnitude, and the average force exerted by the fluid during the passage through the turbine is

$$F_{ave} = (approx)\ M V_{turb}^2/R_v$$

where V_{turb} is the average magnitude of the velocity of the fluid while passing through the turbine.

The above force at any instant is in a direction which is perpendicular to the surface of the turbine at the points of contact between the small

43

volume and the turbine at that instant. Therefore, the component of the force which contributes to the torque on the turbine is less than F_{ave} at all times or substantially at all times. Also, the component of F_{ave} which contributes to the torque contributes only during the time the fluid is in the turbine. The average torque on the turbine resulting from the small volume of fluid making its first complete cycle of movement through the converter, then, is

$$Tq_{turb\ 1} = C_1\ C_2\ M\ (V_{turb\ 1}^2/R_v)\ R_{tq\ arm}$$

In the above, C_2 is a constant with a value less than 1.0 which, when multiplied by the value of F_{ave}, results in a force which is the equivalent of F_{ave} in its torque-producing capability when it is applied to the turbine at a torque-arm radius of $R_{tq\ arm}$. C_1 is a constant that is determined by the dimensions of the converter. It causes the expression to be the equation of a continuous torque acting during the complete cycle which is the equivalent of the temporary torque the small volume produces during its movement through the turbine. $V_{turb\ 1}$ is the average value of the linear velocity of the small volume of fluid during this first passage of the fluid through the turbine.

Now assume that the direction of flow of the small volume is changed by a vane of the stator so that its V_r velocity component is in the forward direction, and so that W_r of the fluid is approximately equal to W_{imp}. Then, it may be possible that the small volume will pass through the impeller without touching a vane of the impeller. Assume that it does so and reenters the turbine with a V_{r4}/V_{c4} ratio, which is substantially the same as existed in its first entry. Assume, further, that it enters along a path which is substantially tangential to the surface of a turbine vane (not necessarily the same vane as in the first passage) at the points of entry into the turbine. Then, the torque produced on the turbine, $Tq_{turb\ 2}$, during the second passage would be substantially the same as during the first, or $Tq_{turb\ 1}$, if the velocity $V_{turb\ 2}$ were not less than $V_{turb\ 1}$. It is less by an amount which is determined by the friction and other forces acting on the fluid during its passage through a complete cycle of movement and during the second passage through the turbine. These forces include those which make $V_{out\ turb}$ slightly less than $V_{in\ turb}$. The torque produced is less than that produced during the first passage of the fluid through the turbine by a factor of approximately $(V_{turb\ 2}/V_{turb\ 1})^2$, where the V values are the average velocities of the fluid in the turbine.

In the second cycle above, substantially no kinetic energy was added by the power source and the impeller since the small volume of fluid did not touch an impeller vane and since the forces acting between the fluid and the impeller wall are assumed to be small. Succeeding cycles in which substantially no kinetic energy is added might be assumed until all of the kinetic energy received by the fluid from the impeller during the first cycle is expended in friction power losses. During each succeeding cycle the fluid would produce a smaller torque on the turbine than during the preceding cycle until the torque produced is zero.

In actual operation of the converter, the assumption that no energy is provided to a small volume of the fluid as it passes through the impeller is the assumption of a condition which cannot be realized during power-on operation since each small volume is in contact with an impeller vane or with a body of fluid which in turn is in contact with a vane. Therefore, energy is added during each passage, but the energy of each small volume as it leaves the impeller is retained, excepting the losses which occur at the times of entering and passing through the turbine and the stator and entering the impeller.

The velocity of each small volume of all of the fluid leaving the turbine is a function of the radius of the volume from the axis of rotation of the converter. Let it be assumed that, in stalled conditions, each of the small-volume velocities is in a direction which is parallel with the surface of the turbine vane against which the volume directly or indirectly is exerting pressure, and that each of these directions also is parallel with the concave surface of the stator vane against which each small volume will exert pressure as it enters the stator. Also assume that the flow of each small volume through the stator is along a path which is the arc of a circle. When each small volume exits from the stator, it has a velocity with a radial component, V_r, and a circulatory component, V_c. These are comparable in magnitude with the values when the fluid left the impeller.

As was assumed earlier, the power source is one which is governed in a manner such that its output power remains constant as the speed changes through wide limits. Then, at a slow speed it produces a large torque and at a high speed it produces a small torque. As stated above, each small volume of fluid enters the impeller from the stator with velocity components which are in directions that correspond with those of the components when the fluid left the impeller, and the magnitudes of the entering components are comparable with those when the volumes left the impeller. Therefore, the torque and power required from the power source to maintain a particular velocity, $V_{out\ imp}$, are much less than they would be if the velocity, V_r, of the fluid entering the impeller were zero, as it is in the case of a fluid clutch when the runner is stalled. Therefore, the speed of the power source increases to a value greater than it would have if the entering V_r were zero in response to the low torque requirement in order to maintain a constant output power. This, in turn, results in a higher velocity of exit of the fluid from the impeller to the turbine, and a higher velocity of reentry into the impeller, and so on. The increases in fluid velocity and in speed of the power source

45

continue until a balance is established between the output power of the source and the power used in the converter as the result of viscous friction, which power increases as the speed of the impeller and the resultant velocity of the fluid increase. The increase in the velocity of the fluid entering the turbine results in greater torque on the turbine. Thus, the increase in turbine output-shaft torque and the decrease in input-shaft torque from the power source cause the ratio, Tq_{out}/Tq_{in}, which is the torque factor of the converter, to increase. In effect, increased power-source speed is exchanged for increased output-shaft torque, as in the case of the gear systems in Chapter 1.

In fluid torque converters of the three-element type,

$$TF_{conv} = Tq_{out}/Tq_{in}$$

usually is approximately 2.0 to 2.2 when the output shaft of the converter is stalled. Higher TF values are possible at the expense of other desirable characteristics.

When the power source is an automobile engine rather than the source above, an increase in the speed of the engine results in an increase in the torque capability of the engine when the speed is less than the maximum-torque speed. Therefore, there may be an increase in the output torque of the converter due to this factor, as well as from the TF of the converter if an automobile engine is used in the case above, but this does not affect the value of the TF of the converter.

When the output shaft is not stalled, the difference between the velocity, W_r, of the fluid entering the turbine and the W velocity of the turbine vanes becomes less as the turbine velocity increases if the impeller velocity remains constant. Also, the centrifugal pump effect of the turbine upon the fluid within it decreases the volume of the fluid that circulates through the converter per second. The result of these changes is a decrease in the output torque and in the torque factor. As the turbine velocity continues to increase, a velocity is reached at which the radial-plane component, W_r, of the velocity of the fluid leaving the turbine and striking the stator vanes is in the forward direction. If the stator remained stationary at this and higher turbine velocities, the stator vanes would reduce the radial-plane component of the velocity of the fluid and thereby require greater torque and power from the engine to accelerate the fluid to the speed of the impeller vanes at the points of entry. Therefore, the stator is connected to an overrunning clutch, described above, which permits it to rotate freely in the forward direction. The stator vanes, then, have little effect on the flow of fluid during its movement between the turbine and the impeller. Under this condition, the converter operates as a fluid clutch with a torque factor that is substantially 1.0 under normal operating conditions.

The point at which the stator starts and stops rotating is the coupling point, or clutch point, and may be identified as the slip in per cent at which the point occurs. The variation of the torque factor with slip is illustrated in Fig. 9.

The total of the input torques placed on the elements of the converter must equal zero since every torque must have an equal and opposing torque. Then,

$$Tq_{in} \text{ in shaft} + Tq_{in} \text{ stator shaft} + Tq_{in} \text{ out shaft} = 0.0$$

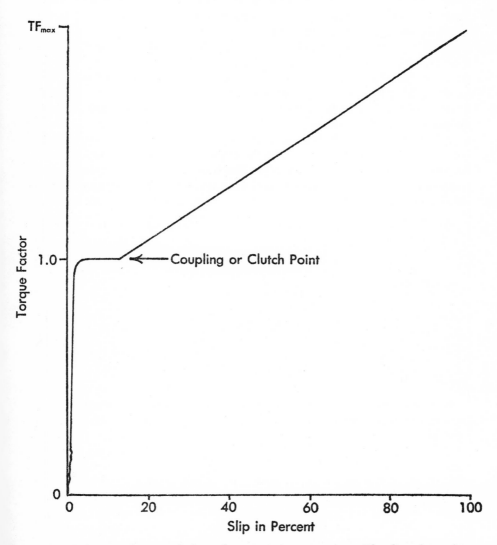

FIG. 9. Torque factors of three-element torque converter. The line from the coupling point to TF_{max} usually is not a straight line as shown. It may vary below or above, or below and above the line.

47

where the Tq_{in} values are provided by the engine, transmission case, and load, respectively. If the output shaft is stalled and the torque factor is 2.2,

$$Tq_{in} \text{ in shaft} - Tq_{out} \text{ st shaft} - Tq_{out} \text{ out shaft} = 0.0, \text{ or}$$
$$Tq_{in} - Tq_{out} \text{ st shaft} = Tq_{out} \text{ out shaft} = 2.2\ Tq_{in}, \text{ and}$$
$$Tq_{out} \text{ st shaft} = Tq_{in} - 2.2\ Tq_{in} = -1.2\ Tq_{in}$$

Thus, the stator places a torque on the transmission case equal to 1.2 times the input torque, but in the opposite direction.

When the slip is small and the torque factor is 1.0,

$$Tq_{out} \text{ st shaft} = Tq_{in} - Tq_{in} = 0.0$$

so the stator places no torque on the transmission case. Instead, it rotates in the forward direction against (substantially) no torque.

As in the case of fluid clutches, the power-on stalled-condition torques of similar torque converters may be expressed as

$$Tq_{out} = TF\ C\ S_{imp}^{2}\ D^5$$

where C and TF are (substantially) the same for all of the similar converters and D is the maximum diameter in feet of the impeller of the converter being considered.

d. POWER-OFF OPERATIONS. In power-off operations, the turbine acts as an impeller and the impeller acts as a turbine, or more as a runner of a fluid clutch, since its vanes resemble those of a runner. The fluid leaving the turbine, or "new" impeller, strikes the stator vanes with a forward velocity. Therefore, the stator rotates freely and the converter operates as a fluid clutch. Normally, its torque factor is substantially 1.0. The items which affect the speed factor and capability of transferring torque and "braking power" to the engine are similar to those discussed in § 5. However, the curvature of the turbine vanes makes the torque capability less than it would be in a fluid clutch of the same size under the same speed conditions.

e. POWER RELATIONSHIPS AND SPEED FACTOR. When the stator rotates freely, the converter acts as a fluid clutch and, from § 5, in power-on operations,

$$P_{out}/P_{in} = S_{out}/S_{in} = SF$$

except when the slip is very small. This relationship applies also to all power-off operations since the stator rotates freely in all of these operations. Likewise, the speed factor in power-on operations has any value from slightly less than 1.0 to a lesser value (approximately 0.85 to 0.90) that exists when the stator ceases rotating. The SF values in power-off operations vary from slightly less than 1.0 to 0.0, as indicated in § 6, d.

When the stator does not rotate, a three-element converter has a TF which varies from substantially 1.0 at the slip value at which the stator ceases rotating to a maximum value which usually is approximately 2.0 to 2.2 when the turbine is stalled. In the following it is assumed that the maximum TF is 2.0.

The ratio of power output to power input is

$$P_{out}/P_{in} = Tq_{out} \, S_{out}/(Tq_{in} \, S_{in})$$
$$= TF \, Tq_{in} \, SF \, S_{in}/(Tq_{in} \, S_{in})$$
$$= TF \, SF$$

When the turbine is stalled, S_{out} and SF are zero. Therefore, $P_{out}/P_{in} = 0.0$. This means that all of the input power from the engine is converted into thermal energy which is dissipated in the form of heat.

When the slip is small, the SF is approximately 1.0, or 0.97 if the slip is 3 per cent. The TF is approximately 1.0—say, 0.99. Then,

$$P_{out}/P_{in} = TF \, SF = (0.99)(0.97) = 0.96$$

Thus, 4 per cent of the output power of the engine is converted into energy that is dissipated in the form of heat.

The ratio, P_{out}/P_{in}, when the slip has values between the above two values, is in the range between 0.0 and the approximate value of 0.96. For example, assume that the slip is 75 per cent and that the TF with this slip is 1.7. Then,

$$P_{out}/P_{in} = (1.7)(0.25) = 0.425$$

and 57.5 per cent of the engine power is dissipated in the form of heat.

The speed factor, as indicated above, varies from slightly less than 1.0 to 0.0, depending on the torque required by the load and the slip required to produce that torque.

f. COMBINED CONVERTER AND AUTOMOBILE CHARACTERISTICS. The combined torque converter and automobile characteristics are similar to those of a fluid clutch and automobile discussed in § 5. The only substantial difference is that due to the difference in torque factors under conditions in which the stator is stationary. It is stationary in power-on operations when the slip exceeds a certain percentage that is determined by the design of the converter. The greater torque factor of the converter permits greater acceleration of the automobile, assuming that other conditions (TF of the transmission gears, etc.) are the same.

Figure 10 illustrates some of the characteristics of an engine-converter combination. The discussion in § 5 concerning Fig. 6 is applicable generally to this illustration.

g. DESIGN FEATURES. In the discussion concerning the torque produced by a torque converter, certain assumptions are made pertaining to the

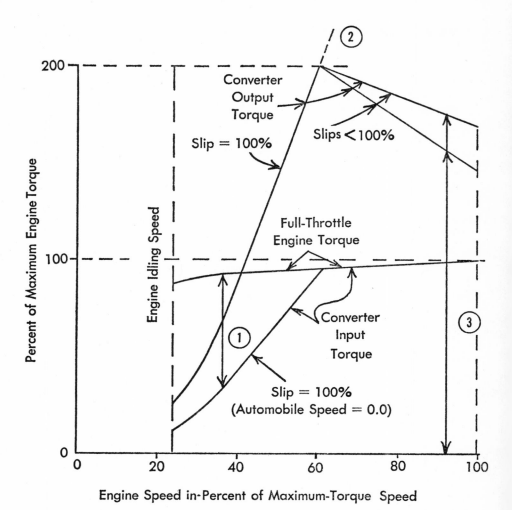

FIG. 10. Three-element torque converter and automobile torques. Low-slip characteristics similar to fluid clutch characteristics. (1) Torque available for engine speed acceleration. (2) Converter output torque if not limited by engine torque. (3) Torque available for existing automobile speed and for acceleration. Converter $TF_{max} =$ Approximately 2.1.

relationships between the directions of the velocity of the fluid and the angles of the surfaces of the impeller, turbine, and stator vanes at the points of entry and exit of the fluid into and out of those elements. However, no consideration is given to the effects on these of the rotation of the turbine, other than the effect which causes the stator to rotate. The turbine rotation, varying in automatic transmissions from zero speed

50

to almost the speed of the impeller in power-on operations, has an effect on the relationships between the fluid directions and the angles of the vanes.

If the angles of the vanes of the elements of the converter with respect to one another are those which produce the best results under one set of conditions, they are not the best under some other conditions. When a converter such as that shown in Fig. 7 is designed, the shapes of the vanes and the angles of the vanes are made to produce the best compromise in satisfying the requirements of the various conditions of operation. One change in the design of the converter in Fig. 7 that is used in some converters permits the angle of the stator vanes to be changed from "low angle" or "low pitch" to "high angle" or "high pitch." When in the high-angle position, the V_r velocity component of the fluid entering the impeller is greater than when in the low-angle position, other factors being constant. The engine speed and the fluid velocity increase as described above. This increases the torque factor of the converter and permits greater torque to be transmitted to the driven wheels of the automobile for use as "pulling power," or for use in acceleration of the automobile. The transmission in Fig. 28 uses a variable-pitch stator for these purposes, and for the purpose of lessening the tendency of the automobile to move when the engine is idling. The closing of the throttle to the idling-speed position, in turn, operates a control mechanism that moves the stator vanes to the high-angle position. This lessens the torque the engine needs to provide the impeller at any given impeller speed. The speed of the engine would increase slightly as the result of the factors discussed in the above pertaining to the velocity of the fluid returning to the impeller if it were possible to change the angle while the engine is idling. However, the speed the engine can attain in this instance is limited by the closed throttle. The considerable increase in fluid velocity that occurs with an open throttle does not occur. Therefore, less kinetic energy is added to the fluid by the engine, less torque is provided the impeller by the engine, and less torque is produced on the turbine than would have been produced with the stator vanes in the low-pitch position.

Converters with more than three elements are used in some transmissions. Their torque and speed characteristics are similar to those of the three-element converter but the maximum torque factor may be considerably greater.

It has been indicated several times in the foregoing that forces of friction exist within a converter. These forces are the result of friction between the "drops" of fluid, as well as some friction within bearings, etc. The power losses due to friction may be great as shown in § 6, e, above. The losses under many conditions of automobile operation are such that it is desirable or necessary to employ water cooling of the transmission fluid in order to limit the fluid temperatures to values that are acceptable.

h. SUMMARY. The torque factors of fluid torque converters used in automobile transmissions are:

TF = approximately 2.0 for three-element converters under power-on and stalled conditions, and considerably greater values for some converters with more than three elements,

TF = approximately 2.0 to substantially 1.0 for three-element converters, and greater than 2.0 to substantially 1.0 for some converters with more than three elements, under normal power-on and nonstalled conditions, the 1.0 values pertaining to slip conditions under which the stator (or stators) rotates freely, that is, the lesser slip values,

TF = substantially 1.0 under normal power-off conditions, in which the power flow is from the turbine shaft to the impeller shaft.

The speed factor varies from 0.0 to slightly less than 1.0 in power-on or power-off operations depending on the torque required by the load and the slip required to produce that torque.

The power relationship of a converter is

$$P_{out}/P_{in} = \text{Power Efficiency} = SF\ TF$$

which varies from 0.0 to slightly less than 1.0 as the slip varies from 100 per cent to a few per cent. At some lesser slip, it again becomes 0.0 since the output torque becomes zero with this and smaller slips. The input power that is not transferred to the output shaft must be dissipated in the form of heat, which may require continuous circulation of the converter fluid from and to its housing in order to provide the dissipation at a rate which prevents excessive fluid temperatures.

The output torque of a converter is affected by its dimensions, the viscosity of the fluid and other such factors in ways that are similar to those applicable to fluid clutches, as well as by the greater torque factor of the converter under some conditions. For any given impeller speed it is greatest under stalled conditions, and decreases to 0.0 when the slip has some value near zero slip. With this slip all of the input torque is used in providing the torque required within the converter as the result of friction forces.

3

Planetary Gear Sets and Systems

1. Planetary Gear Sets

Planetary gear sets are illustrated in Figs. 11, 17, 19, 21, and 23–33. They usually are of a sun-and-planet type in which one or more sets of planet gears may rotate around their own axes and around the axis of the sun gear or sun gears. In many instances, a ring gear is employed in the gear set. These gears have internal rather than external teeth, and therefore are known also as internal gears. In some instances no sun gear is used.

Planetary gear sets are known also as epicyclic gear sets since a point on a tooth of a planet gear moves along a path which is an epicycloid line or path.

For the purposes of this discussion, the gear sets are classified as simple or compound sets. Simple gear sets are those which have only three elements (sun gears, planet carriers, and ring gears) that may be connected to external devices by shafts or other connecting links. Compound gear sets have more than three such elements. Figures 11, 19, 21, 24, 27, 29, 30, 32, and 33 include one or more simple sets, and Figs. 17, 23, 25, 28, and 31 have one compound set. The set shown in Fig. 26 might be classified as a split simple set since the shafts of sun gears 1 and 2 rotate at the same speed and pinions 1 and 2 rotate at the same speed.

The planet pinions of a set of pinions are identical and perform the same functions. There may be one, two, three, or more pinions in a set, but the usual numbers are three and four. The speed and the torque factors of the gear set are the same, whatever the number of pinions, when the numbers of teeth on the different gears remain the same. The computations of torque factors sometimes are simpler if the gear set is assumed to have only one pinion in each set of pinions. This assumption is used in some instances in following chapters. The use of more than one pinion in a set reduces the forces acting on a tooth of the gears, provides a better balance of the centrifugal forces which result from rotation of the planet carrier, and eliminates or reduces greatly some of the forces that exist between the shafts and their bearings.

2. Planetary Gear Systems

a. SYSTEM ELEMENTS. A planetary gear system includes a planetary gear set or sets and members which connect the elements of the gear set or sets to shafts or to other devices, exclusive of the planet pinion elements but including the planet carrier or carriers. In Figs. 11, 17, 19, and 21 all of the elements (except planet pinions) are connected to shafts. They are illustrated in this way in order to permit the use of the figures in the studies of systems in different conditions of operation (see Appendix A, Definitions). In practice, one or more of the elements may be connected permanently to a stationary structure, or to a brake drum which is permitted to rotate or is stopped by a brake band depending on the desired result, or used in some other manner which does not require the connection of the element to a shaft. Likewise, one or more shafts or gear set elements may be designated as the input shaft(s) or element(s) and one or more as the output shaft(s) or element(s). Figures 23–33 illustrate some of the element connections that are used in transmissions. In the following definitions it is assumed that all elements of the gear sets are connected to shafts, exclusive of the planet pinions.

b. SIMPLE PLANETARY GEAR SYSTEMS. Simple planetary gear systems employ one planetary gear set and have only one input shaft and one output shaft. A source of power is connected to the input shaft and a load mechanism is connected to the output shaft. These systems may employ simple or compound gear sets.

c. COMPOUND PLANETARY GEAR SYSTEMS. These systems employ one simple or compound gear set and more than one input shaft or more than one output shaft, or more than one of each.

d. DUAL, ETC., PLANETARY GEAR SYSTEMS. Dual, triple, etc., systems employ more than one planetary gear set connected in a way which causes the output shaft or shafts of one to provide torque to the input shaft or shafts of another. Figures 24, 29, 32, and 33 include dual or triple systems. In these illustrations some of the connecting links between the elements of the gear sets are not shafts. The entire dual, triple, etc., gear system may have only one input shaft and one output shaft. Therefore, it has the torque and speed characteristics of a simple gear system discussed in Chapter 1, and in that sense is a simple gear system.

Simple Planetary Gear Systems Employing Simple Gear Set With Ring Gear

1. Numbers of Gear Teeth

From inspection of Fig. 11, the relationship between the diameters of the gears (see Appendix A, Definitions) is

$$D_r = D_s + 2D_p$$

where subscripts r, s, and p, indicate the ring, sun, and planet gears.

The number of teeth on a gear is

$$T = D\pi/T_{ft} = D\,C$$

where $C = \pi/T_{ft}$, D is in feet, and T_{ft} is the number of teeth per foot of circumference of the gear. Then,

$$D = T/C$$

Assuming that C is the same value for the ring and sun gears and planet pinions, the diameter equation may be written as

$$T_r/C = T_s/C + 2T_p/C, \text{ or}$$
$$T_r = T_s + 2T_p, \text{ or}$$
$$T_p = (T_r - T_s)/2$$

The T values must be whole numbers, and $(T_r - T_s)$ must be an even number in order that T_p will be a whole number. Thus, each of T_s and T_r must be an even number or each must be an odd number if standard proportions of height of gear tooth above the pitch circle and depth of the gear tooth below the pitch circle are used. Other proportions used in planetary gear sets permit a departure from this odd and even number relationship. In this and following chapters pertaining to torque and

55

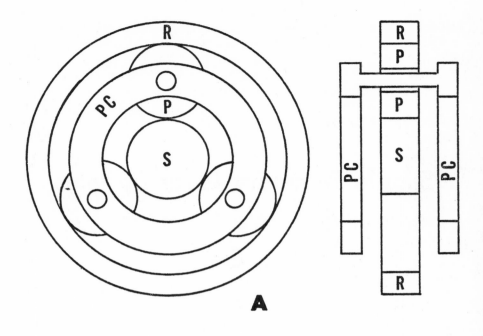

A

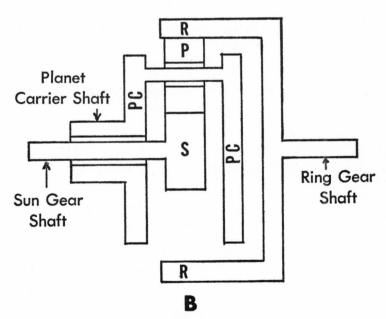

Planet
Carrier Shaft

Sun Gear
Shaft

Ring Gear
Shaft

B

Fig. 11. Simple planetary gear set with ring gear.

speed factors of gear systems it is assumed that standard tooth proportions are used. In chapters pertaining to transmissions, standard or nonstandard proportions may be assumed to exist.

It can be shown that the following relationships between the numbers of teeth on the sun gear and planet pinions must exist if the pinions are to be spaced equally around the sun gear.

Sets with Three Planet Pinions:

(1) If T_s is divisible by 3, T_p must be divisible by 3.

(2) If $(T_s - 1)$ is divisible by 3, $(T_p + 1)$ must be divisible by 3.

(3) If $(T_s + 1)$ is divisible by 3, $(T_p - 1)$ must be divisible by 3.

Sets with Four Planet Pinions:

(1) If T_s is even, T_p must be even.

(2) If T_s is odd, T_p must be odd.

2. Conditions of Operation

Any Condition of Operation includes at least one input shaft and at least one output shaft. In simple gear systems, there is only one of each of these shafts (see Appendix A, Definitions).

There are 18 possible conditions of operation of simple gear systems employing the gear set shown in Fig. 11. The shafts used as an input shaft, etc., in each of the conditions are shown in Chart 1.

a. CONDITIONS 1–6. In each of these conditions, one of the shafts in Fig. 11 is used as an input shaft, another as the output shaft, and the third is Held, which means held stationary.

b. CONDITIONS 7–12. In these conditions, the input and output shafts are the same as in Conditions 1–6 but the third shaft is Free, which means free to rotate with (substantially) no torque opposing its rotation.

c. CONDITIONS 13–15. Two of the shafts are locked together, directly or indirectly, and one of them is used as the input shaft. The third shaft is the output shaft.

d. CONDITIONS 16–18. These conditions are the inverse of Conditions 13–15 in that one of the locked shafts is used as the output shaft and the third shaft is the input shaft.

3. Condition 1

a. SHAFT CONDITIONS (Fig. 11):

Sun gear shaft   —   Input
Planet carrier shaft   —   Held
Ring gear shaft   —   Output

b. SPEED FACTOR. Each revolution of the input shaft and sun gear in the forward direction causes the planet pinions to make T_s/T_p revolutions around their axes in the reverse direction (see Appendix A, Definitions).

Each revolution of the planet pinions causes the ring gear and its shaft to turn T_p/T_r revolutions in the same direction, or reverse direction. Therefore, each revolution of the input shaft causes

$$(T_s/T_p)\ (T_p/T_r) = T_s/T_r$$

revolutions of the output shaft in the reverse direction, or $-T_s/T_r$ revolutions, where the $-$ sign indicates the reverse direction with respect to the direction of rotation of the input shaft.

Since all of the above revolutions occur in the same time period, each revolution per minute (rpm) of the input shaft in the forward direction results in $-T_s/T_r$ rpm of the output shaft.

The speed factor is

$$SF = S_{out}/S_{in} = (-T_s/T_r)\ (S_{in})/(S_{in}) = -T_s/T_r$$

c. Torque Factor. The torque factor in a simple gear system is the reciprocal of the speed factor (Chapter 1, § 1), so

$$TF = Tq_{out}/Tq_{in} = -T_r/T_s$$

4. Condition 2

a. Shaft Conditions (Fig. 11):

Sun gear shaft   —   Output
Planet carrier shaft —   Held
Ring gear shaft   —   Input

b. Speed and Torque Factors. This condition is the inverse of Condition 1. Therefore (Chapter 1, § 1),

$$SF = -T_r/T_s, \text{ and}$$
$$TF = -T_s/T_r$$

5. Condition 3

a. Shaft Conditions (Fig. 11):

Sun gear shaft   —   Held
Planet carrier shaft —   Input
Ring gear shaft   —   Output

b. Speed and Torque Factors. When the input shaft rotates, the resultant rotation of the ring gear and the output shaft has two components: (1) The rotation that would exist if the planet pinions did not rotate on their axes but slid along the surface of the sun gear, and (2) the rotation of the ring gear that exists as the result of the planet pinions rotating on their axes.

When the input shaft and the planet carrier make one revolution, the

58

first component of the rotation of the ring gear and the output shaft also is one revolution in the forward direction.

When the input shaft and the planet carrier make one revolution, each planet pinion moves around the sun gear and in so doing makes T_s/T_p revolutions about its axis. Each of the revolutions made by the set of pinions causes the ring gear to make T_p/T_r revolutions about its axis in the forward direction. The second-component revolutions of the ring gear, then, are

$$(T_s/T_p) \, (T_p/T_r) = T_s/T_r \text{ revolutions}$$

in the forward direction.

The total number of revolutions of the ring gear and the output shaft which results from one revolution of the input shaft is the sum of the two components, or $1 + T_s/T_r$ revolutions in the forward direction, so

$$SF = 1 + T_s/T_r, \text{ and}$$
$$TF = 1/SF = 1/(1 + T_s/T_r)$$

6. Condition 4

a. SHAFT CONDITIONS (Fig. 11):

Sun gear shaft	—	Held
Planet carrier shaft	—	Output
Ring gear shaft	—	Input

b. SPEED AND TORQUE FACTORS. This condition is the inverse of Condition 3. Therefore,

$$SF = 1/(1 + T_s/T_r), \text{ and}$$
$$TF = 1 + T_s/T_r$$

7. Condition 5

a. SHAFT CONDITIONS (Fig. 11):

Sun gear shaft	—	Input
Planet carrier shaft	—	Output
Ring gear shaft	—	Held

b. SPEED AND TORQUE FACTORS. A pound of force at the end of a 1 ft torque arm connected to the input shaft (1 lb-ft of torque) causes a force of $(1) \, (1/R_s)$ lb in the forward direction at the points of contact between the teeth of the sun gear and those of the planet pinion (see Appendix A, Definitions, gear radius). This force, in fact, is divided between the points of contact of the three pinions and the sun gear when there are three pinions but, as stated in Chapter 3, § 1, it is assumed for purposes of computation that only one pinion exists in the gear set.

59

Each pound of force at the junction of the sun gear and the planet pinion causes 2 lb of force at the axis of the pinion and 2 lb of force on the planet carrier at the radius of $R_s + R_p$. This force is in the forward direction.

The force at the end of a 1-ft torque arm attached to the output shaft as the result of 1 lb of force on the input shaft torque arm (1 lb-ft of torque) is

$$(1)\ (1/R_s)\ (2)\ (R_s + R_p)/(1) = (2/R_s)\ (R_s + R_p)\ \text{lb}$$

Since $R_p = (R_r - R_s)/2$, the output shaft torque arm force and torque are

$$(2/R_s)\ (R_s + [R_r - R_s]/2) = (1 + R_r/R_s)\ \text{lb and lb-ft}$$

Since

$$D_r = 2R_r = T_r/C, \text{ and } D_s = 2R_s = T_s/C$$

where $C = \pi/T_{ft}$, the torque on the output shaft is $1 + T_r/T_s$ lb-ft in the forward direction, and this torque results from 1 lb-ft of torque that is applied to the input shaft. Therefore,

$$TF = 1 + T_r/T_s, \text{ and}$$
$$SF = 1/TF = 1/(1 + T_r/T_s)$$

An analysis of the rotations of the various elements for the purpose of determining the speed factor may be made by either one of two methods. The method in which one rotation of the input shaft is assumed and a determination is made of the resultant number of output shaft revolutions involves more difficult computations than in the preceding conditions. In the second and easier method, one revolution of the output shaft is assumed and a determination is made of the number of input shaft revolutions required to produce the output shaft revolution. This, in effect, is a determination of the speed factor in Condition 6.

8. Condition 6

a. SHAFT CONDITIONS (Fig. 11):

Sun gear shaft — Output
Planet carrier shaft — Input
Ring gear shaft — Held

b. SPEED AND TORQUE FACTORS. This condition is the inverse of Condition 5, so

$$SF = 1 + T_r/T_s, \text{ and}$$
$$TF = 1/(1 + T_r/T_s)$$

9. Condition 7

a. Shaft Conditions (Fig. 11):

Sun gear shaft	—	Input
Planet carrier shaft	—	Free
Ring gear shaft	—	Output

b. Speed and Torque Factors. This condition is similar to Condition 5 if the load on the output shaft in Condition 7 is considered as tending to Hold that shaft. The force on the planet carrier is $(2)(1/R_s) = 2/R_s$ times the force acting on a 1 ft torque arm attached to the input shaft. Since the planet carrier is Free to rotate against no force except that of friction, the above force and the torque on the ring gear and the output shaft are (substantially) zero. Therefore, the TF may be considered to be zero (or negligible) and the output shaft does not rotate if there is an appreciable load connected to it. Therefore, the SF is zero.

10. Conditions 8–12

a. Shaft Conditions (Fig. 11):

Any shaft	—	Input
Any other shaft	—	Output
Remaining shaft	—	Free

This listing includes Condition 7 as well as Conditions 8–12.

b. Speed and Torque Factors. Reasoning such as that for Condition 7 shows that no torque (or negligible torque) is transmitted by the systems and the output shafts do not rotate if there is an appreciable load on them. Therefore,

$$SF = 0.0, \text{ and}$$
$$TF = 0.0$$

11. Condition 13

a. Shaft Conditions (Fig. 11):

Sun gear and planet carrier shafts	—	Input and locked together
Ring gear shaft	—	Output

b. Speed and Torque Factors. When the sun gear and the planet carrier shafts are locked together, directly or indirectly, they rotate together and the planet pinions do not rotate about their axes. Therefore, the ring gear and the output shaft rotate at the same speed as the input shaft and in the forward direction. The SF and the TF, then, are 1.0.

12. Conditions 14–18

a. SHAFT CONDITIONS (Fig. 11):

Any two shafts   —   Input or output, and locked together
Remaining shaft   —   Output or input

This listing includes Condition 13 as well as Conditions 14–18.

b. SPEED AND TORQUE FACTORS. Reasoning such as that for Condition 13 shows that the SF and the TF for each of these conditions are 1.0.

13. Summary and Graphs

The speed and torque factors determined in the above are indicated in Chart 1. In Appendix C is a discussion of a nonplanetary gear system which is the equivalent of the planetary gear set shown in Fig. 11.

The speed and torque factors in Conditions 1–6 are of three general forms. These are shown in Table 4.

TABLE 4. SPEED AND TORQUE FACTORS IN CONDITIONS 1–6

Form No.	General Form	Corresponding TF or SF Values
1	$-a/b$	$-T_s/T_r$ and $-T_r/T_s$
2	$1+a/b$	$1+T_s/T_r$ and $1+T_r/T_s$
3	$1/(1+a/b)$	$1/(1+T_s/T_r)$ and $1/(1+T_r/T_s)$

In some instances, in Table 4, a represents T_s and in others it represents T_r. Likewise, b represents one or the other of these values. Also a may be associated with a positive or a negative sign. When a represents T_s, the ratio a/b must be less than 1.0; and when it represents T_r, the ratio must be greater than 1.0.

The torque factors in compound systems employing the gear set shown in Fig. 11, and some of the speed and torque factors in simple systems which employ the compound planetary gear set shown in Fig. 17, have forms which are the same as one or another of the above three forms. Graphs of the values of the three forms versus values of a/b then can serve as graphs of the SF or TF values versus values of a/b for any gear set in any condition of operation in which the SF or TF has a value corresponding with one of these forms. For this reason, the SF and TF values are plotted in Figs. 12 and 13 versus a/b rather than versus T_s/T_r or T_r/T_s. Chart 2 provides information concerning the applicability of Figs. 12 and 13 and Forms 1, 2, and 3 to the various conditions of operation of simple and compound planetary gear systems.

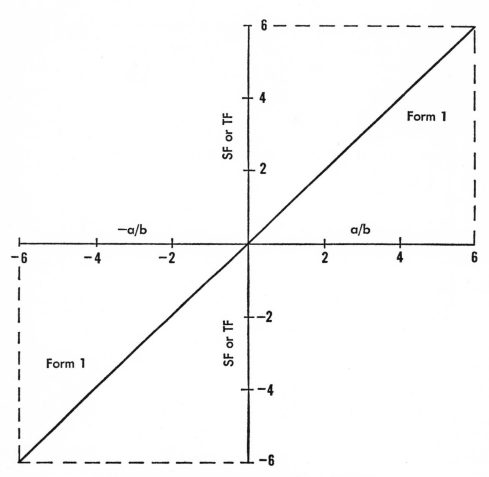

CHART 1. SIMPLE PLANETARY GEAR SYSTEMS
(Simple Gear Set—Fig. 11).

Shaft				SF	Graph	
Cond.	S	PC	R		Fig.	Form
1	I	H	O	$-T_s/T_r$	12	1
2	O	H	I	$-T_r/T_s$	12	1
3	H	I	O	$1 + T_s/T_r$	13	2
4	H	O	I	$1/(1 + T_s/T_r)$	13	3
5	I	O	H	$1/(1 + T_r/T_s)$	13	3
6	O	I	H	$1 + T_r/T_s$	13	2

FIG. 12. Speed and torque factors—Form 1. (1) Form 1—SF or $TF = \pm a/b$. (2) Chart 2 states applicability.

CHART 1. SIMPLE PLANETARY GEAR SYSTEMS (*cont.*)

Conditions 7–12

Any shaft	—	Input
Any other shaft	—	Output
Remaining shaft	—	Free

$SF = 0.0 \qquad TF = 0.0$

Conditions 13–18

Any two shafts locked together	—	Input (or Output)
Remaining shaft	—	Output (or Input)

$SF = 1.0 \qquad TF = 1.0$

Notes: (1) $SF = S_{out}/S_{in}$

(2) $TF = Tq_{out}/Tq_{in} = 1/SF$ except in Conditions 7–12

(3) I, Input; O, Output; H, Held (stationary); Free, Free to rotate

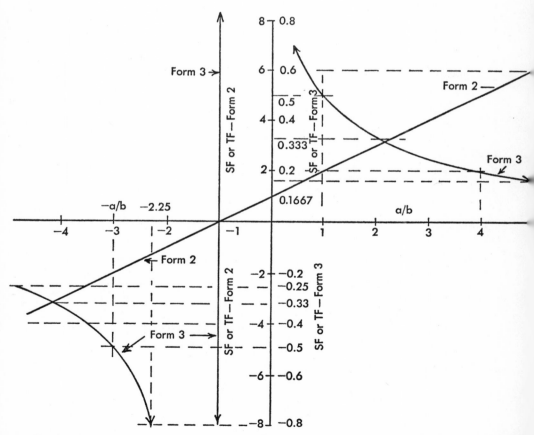

FIG. 13. Speed and torque factors—Forms 2 and 3. (1) Form 2—*SF* or $TF = 1 \pm a/b$. Form 3—*SF* or $TF = 1/(1 \pm a/b)$. (2) Chart 2 states applicability.

CHART 2. APPLICABILITY OF FIGS. 12 AND 13

Fig.	Form	Simple Systems				Compound Systems	
		SF		TF		TF$_1$	TF$_2$
		Fig. 11	Fig. 17	Fig. 11	Fig. 17	Fig. 11	Fig. 11
12	1	Cond. 1, 2	Cond. 1, 2, 7, 8, 23, 24	Cond. 1, 2	Cond. 1, 2, 7, 8, 23, 24	Cond. 1, 2	Cond. 3, 4
13	2	3, 6	3, 6, 9, 15, 18, 21	4, 5	4, 5, 10, 16, 17, 22	4, 5	2, 5
13	3	4, 5	4, 5, 10, 16, 17, 22	3, 6	3, 6, 9, 15, 18, 21	3, 6	1, 6

5

Compound Planetary Gear Systems Employing Simple Gear Set With Ring Gear

1. Conditions of Operation

There are six possible conditions of operation of compound gear systems employing the gear set illustrated in Fig. 11. These are listed in Chart 3 in which I_1 and I_2 are input shafts 1 and 2 and O_1 and O_2 are output shafts 1 and 2.

It is possible that the load connected to an output shaft, or to each of two output shafts, will provide power to a system for a period of time rather than obtain power from the system. In that event, the condition of operation changes from one to another. For example, assume that Condition 2 exists when both loads connected to the system receive power from the system. If the load connected to output shaft 1 provides power to the system for a time, then another condition exists during that time. Therefore, the possibility of one or two loads providing power to a system under any one of Conditions 1–6 is not considered in the discussion of that condition since the discussion of the new condition that exists applies.

Likewise, it is possible that a power source connected to one of the input shafts in Condition 1, 3, or 5 will be driven by the power source connected to the other input shaft, and thus act as a load on the gear system instead of as a power source for the system. In that event, Condition 2, 4, or 6 exists rather than Condition 1, 3, or 5, and the discussion of the new condition applies.

Chapter 1, § 2 discusses the speed and torque factors that pertain to compound planetary gear systems which employ the gear set shown in Fig. 11.

66

2. Torque Relationships

Assume the following: (1) The gear set shown in Fig. 11 is supported by conical pivots which fit in conical bearings located on the axis of the gear set, one bearing in the end of the sun gear shaft and the other in the end of the ring gear shaft. (2) Two identical 1-ft torque arms are attached to each shaft in horizontal positions so that their ends are 1 ft from and on opposite sides of the axis of their shaft. (3) Weights may be hung on the ends of the torque arms, or the ends of these arms may be attached to stationary objects.

If a planet carrier torque arm is connected to a stationary object and a weight of P lb is hung on the end of a sun gear torque arm, this weight can produce a torque on that arm of P lb-ft in the direction which is assumed in the following to be the forward direction. This torque (Condition 1, Chart 1) can produce a torque on the torque arms attached to the ring gear shaft of $(-T_r/T_s)$ (P) lb-ft, which tends to turn that shaft in the reverse direction. However, if a weight of (T_r/T_s) (P) lb is hung on a ring gear torque arm so as to tend to turn the ring gear shaft in the forward direction, the weights on the sun and the ring gear torque arms produce equal and opposite torques and the sun and the ring gear shafts remain stationary.

The torque on the sun gear shaft produces a torque on the planet carrier shaft (Chapter 4, § 7) of $(1 + T_r/T_s)$ (P) lb-ft in producing the above torque on the ring gear shaft. The force on the end of the planet carrier torque arm that acts on the stationary object, then, is $(1 + T_r/T_s)$ (P) lb-ft in the forward direction, and the force the stationary object exerts on the end of the torque arm is the same but it is in the reverse direction. Therefore, all of the shafts remain stationary with respect to one another if a weight of $(1 + T_r/T_s)$ (P) is hung on the end of a planet carrier torque arm so as to tend to rotate that arm in the reverse direction and the connection to the stationary object is removed. Likewise, the gear set as a whole remains stationary since the torque acting on it in the reverse direction is that on the planet carrier shaft of $(1 + T_r/T_s)$ (P) lb-ft and that acting in the forward direction is the sum of the torques on the sun and ring gear shafts which is $(1 + T_r/T_s)$ (P) lb-ft.

The above relationships between the torques exist when the shafts rotate at a constant speed as well as when stationary, assuming the torques required within the gear system as the result of friction to be negligible. Likewise, they exist when the speeds of rotation are changing, assuming the inertia of each of the gear system parts to be negligible in comparison with those of the mechanisms connected to the shafts. In these instances, the torques involved in the (positive or negative) acceleration of the mechanisms attached to the shafts are components of the torques on the shafts. For example, assume that a load connected to a shaft requires X lb-ft of torque at its rotational speed when that speed is constant. Assume further that the output torque of the gear system is increased to a constant value of $(X + Y)$ lb-ft. The difference between these torques, Y lb-ft, increases the speed of the load, and the acceleration of the load at each instant is such that the total torque presented by the load, including that due to the acceleration, is $(X + Y)$ lb-ft. The acceleration ceases when the speed of the load is such that the torque required by the load at the speed attained is $(X + Y)$ lb-ft.

The torques produced by the weights in the above, or their equivalents produced by mechanisms, are input torques. The equivalents may be input torques produced by power sources, input torques presented by loads, or input torques provided by stationary objects. Then,

$$Tq_{\text{in r}} + Tq_{\text{in s}} + Tq_{\text{in pc}} = 0.0$$

The output torque to a load attached to a shaft is equal to the torque provided by the load in magnitude, but it is of the opposite sign. For example, if the torque produced by a force of P lb on the end of the 1-ft sun gear torque arm is the input torque from a power source, the output torques produced by this torque on loads connected to the other shafts are

$$Tq_{\text{out r}} = -Tq_{\text{load r}} = (-T_r/T_s)(P)(1) = Tq_{\text{in}}(-T_r/T_s), \text{ and}$$
$$Tq_{\text{out pc}} = -Tq_{\text{load pc}} = (1 + T_r/T_s)(P)(1) = Tq_{\text{in}}(1 + T_r/T_s)$$

since the input shaft torque arm is 1 ft long.

The torque factors, then, are

$$TF \text{ pertaining to ring gear} = Tq_{\text{out r}}/Tq_{\text{in}} = -T_r/T_s, \text{ and}$$
$$TF \text{ pertaining to planet carrier} = Tq_{\text{out pc}}/Tq_{\text{in}} = 1 + T_r/T_s$$

which are the torque factors in simple system Conditions 1 and 5 in Chart 1.

The ratio of the torques of any pair of shafts of a given gear set, then, has a constant value that is determined by the T_r/T_s ratio. The sign of any particular ratio is positive or negative, depending on whether the torques involved are the torques that are applied to the shafts by the power sources, loads or stationary object, or the torques applied to these by the shafts, or a combination of these torques. Thus, the ratio of the torques applied to the loads of any particular compound system which has two output shafts and employs a gear set of the type shown in Fig. 11 is a constant that is determined by the T_r/T_s ratio of the gear set. Likewise, the ratio of the torques that are applied to the two input shafts of any particular compound system is a constant that is determined by the T_r/T_s ratio of the gear set. Since

$$Tq_{\text{in r}} + Tq_{\text{in s}} + Tq_{\text{in pc}} = 0.0,$$
$$Tq_{\text{out r}} = Tq_{\text{in s}} + Tq_{\text{in pc}}, \text{ and}$$
$$Tq_{\text{in s}} = Tq_{\text{out r}} + Tq_{\text{out pc}}, \text{ etc.}$$

Thus, the sum of the output torques equals the sum of the input torques in compound systems (and in simple systems) when the signs of the torques are used in the addition, so the TF of a compound system might be considered to be 1.0. Simple system Conditions 13–18 in Chart 1 in which the torque factors are 1.0 may be considered as applying to compound systems of special types.

Assume that the sun and the ring gear shafts of a gear set are used as input and output shafts and the planet carrier is Held by a band and drum or by a clutch to provide a Reverse gear in a transmission. Then the torque placed on the Holding device by the planet carrier is

$$Tq_{\text{out pc}} = Tq_{\text{in s}} - Tq_{\text{out r}}$$
$$= Tq_{\text{in s}} - TF\ Tq_{\text{in s}}$$
$$= Tq_{\text{in}}\ (1 - TF)$$

Since TF in this case is negative, the band and drum or the clutch must Hold the planet carrier when the torque acting on the junction between the band and drum or clutch plates is $(1 + TF)\ Tq_{\text{in}}$, where the sign of the TF is considered to be positive. Thus, if T_r/T_s is 3.0, the TF in Reverse gear is —3.0 and the torque on the Holding device and on the transmission case, to which the device is connected, is 4.0 Tq_{in}, or 1,000 lb-ft if Tq_{in} is 250 lb-ft.

If the above gear set is used with the planet carrier as the output element and with the ring gear Held to provide the Low gear of the transmission, the torque factor is 4.0, and

$$Tq_{\text{out r}} = Tq_{\text{in s}} - Tq_{\text{out pc}}$$
$$= Tq_{\text{in}} - 4.0\ Tq_{\text{in}} = -3.0\ Tq_{\text{in}}$$

The Holding device and the transmission case, then, are subjected to a torque of —3.0 Tq_{in}, or —750 lb-ft if Tq_{in} is 250 lb-ft.

The principles discussed in this section apply also to systems which employ other planetary gear sets. However, as indicated in Chapter 7, the natures of the torque relationships may not be the same as those in the gear set shown in Fig. 11.

3. Condition 1

a. SHAFT CONDITIONS (Fig. 11):

Sun gear shaft	—	Input 1
Planet carrier shaft	—	Input 2
Ring gear shaft	—	Output

This condition is a combination of Conditions 1 and 3 of the simple gear systems. These conditions (Chart 1) are:

	Simple System	
Shaft or Factor	Condition 1	Condition 3
Sun gear	Input	Held
Planet carrier	Held	Input
Ring gear	Output	Output
Speed factor	$-T_s/T_r$	$1 + T_s/T_r$
Torque factor	$-T_r/T_s$	$1/(1 + T_s/T_r)$

The direction of rotation of input shaft 1, and the direction of the torque applied to that shaft by its power source, is the forward direction (see Appendix A, Definitions).

b. SPEED FACTORS. When input shafts 1 and 2 rotate at speeds of $S_{in\ 1}$ and $\pm S_{in\ 2}$, the speed of the output shaft is the sum of the components that are produced by each of the input shafts, or

$$S_{out} = (S_{in\ 1})\ (-T_s/T_r) + (\pm S_{in\ 2})\ (1 + T_s/T_r)$$

The direction of rotation of the output shaft is forward if the sum of the two components is positive and reverse if it is negative.

The general definition of the speed factor is $SF = S_{out}/S_{in}$, but in this case there are two input shaft speeds. Therefore, as discussed in Chapter 1, § 2, two speed factors are used

$$SF_1 = S_{out}/S_{in\ 1},\ \text{and}$$
$$SF_2 = S_{out}/S_{in\ 2}$$

in which these two factors apply to a specific value of A, and

$$A = \pm S_{in\ 2}/S_{in\ 1}$$

Then, $S_{in\ 2} = A\ S_{in\ 1}$. Substituting this value of $S_{in\ 2}$ in the above equation and dividing by $S_{in\ 1}$,

$$SF_1 = S_{out}/S_{in\ 1} = -T_s/T_r + A(1 + T_s/T_r),\ \text{or}$$

$SF_1 = (T_s/T_r)\ (A - 1) + A$, for a specified value of A. As discussed in Chapter 1, § 2, $SF_2 = SF_1/A$, so

$$SF_2 = S_{out}/S_{in\ 2} = (T_s/T_r)\ (1 - 1/A) + 1$$

for a specified value of A.

If $S_{in\ 1}$ is zero and $S_{in\ 1}$ is not zero, then A is zero and $SF_1 = -T_s/T_r$, which is the same as the SF in Condition 1 of the simple system. Likewise, if $S_{in\,1}$ is zero and A is infinite, $SF_2 = 1 + T_s/T_r$, which is the same as the SF in Condition 3 of the simple system.

As shown below, A must be a negative value when Condition 1 exists. However, the above speed factors apply whether or not A is a negative value.

c. Torque Factors. From the discussion in § 2, the torque factors when Condition 1 exists (see below) are:

$$TF_1 = Tq_{out}/Tq_{in\,1} = -T_r/T_s, \text{ and}$$
$$TF_2 = Tq_{out}/Tq_{in\,2} = 1/(1 + T_s/T_r), \text{ and}$$
$$U = \pm Tq_{in\,2}/Tq_{in\,1} = TF_1/TF_2 = -(1 + T_r/T_s)$$

The direction of rotation of input shaft 1, by definition, is the forward direction. The torque provided by the power source connected to this shaft, then, must be in the forward direction. Since U is negative, the torque provided by the power source connected to input shaft 2 is in the reverse direction. If shaft 2 is to serve as an input shaft, the direction of rotation of the power source connected to it must be in the direction of its torque or in the reverse direction. Then,

$$A = \pm S_{in\,2}/S_{in\,1}$$

must be negative in Condition 1.

As indicated above, some of the TF_1, TF_2, and U values do not apply in some instances in which it may appear that Condition 1 exists, when TF_1, etc., are based on the torques produced by the power sources when acting as power sources, rather than as loads. The reason for departure from Condition 1, and therefore from these values, is the relationship that may exist between the maximum torque capability of one power source and that of the other. As an example of these departures, assume the following:

(1) The T_r/T_s ratio of the gear set is 2.0.

(2) The directions of $Tq_{in\,1}$ and $Tq_{in\,2}$ are forward and reverse, respectively.

(3) The maximum torque capabilities of the power sources are: power source 1, 10 lb-ft; power source 2, 12 lb-ft.

(4) The torque required by the load in lb-ft, which equals the required Tq_{out}, varies with the speed in rpm of the load input shaft, or S_{out} of the gear system, in accordance with the expression,

$$Tq_{out} = C\,S_{out}^2$$

where C is a constant.

(5) The torque provided by power source 1 to input shaft 1 is increased in steps of 1 lb-ft from 0.0 to its maximum capability of 10 lb-ft.

Then, as the torque of power source 1 is increased, the torque pro-

71

vided by power source 2, $Tq_{\text{in }2}$, and other items are as indicated in the partial listing of values shown in Table 5.

TABLE 5. TORQUE RELATIONSHIPS

$Tq_{\text{in }1}$	Tq_{out}	$Tq_{\text{in }2}$	U	$Tq_{\text{in }2} + Tq_{\text{in }1}(1 + T_r/T_s)$
1	-2	-3	-3.0	$-3 + 3 = 0$
4	-8	-12	-3.0	$-12 + 12 = 0$
5	-10	-12	-2.4	$-12 + 15 = 3$
8 -	-16	-12	-1.5	$-12 + 24 = 12$
10	-20	-12	-1.2	$-12 + 30 = 18$

Table 5 shows that the TF_1, TF_2, and U values are as stated above until the torque capability of power source 2 becomes less in magnitude than the torque placed on its shaft by power source 1 through the gear set. When this condition exists, as indicated by the nonzero values in the last column, power source 2 is driven by source 1 in a direction opposite to its normal direction of rotation (assuming that it can be driven in that direction) and therefore becomes a load. The positive nonzero values of the resultant torques shown in the last column are balanced by the load torques provided by power source 2. Condition 1, then, no longer exists, and Condition 4 of this Chapter exists.

4. Condition 2

a. SHAFT CONDITIONS (Fig. 11):

Sun gear shaft   —   Output 1
Planet carrier shaft   —   Output 2
Ring gear shaft   —   Input

This is the inverse of Condition 1 and is a combination of simple-system Conditions 2 and 4 in Chart 1. These conditions are:

	Simple System	
Shaft or Factor	Condition 2	Condition 4
Sun gear	Output	Held
Planet carrier	Held	Output
Ring gear	Input	Input
Speed factor	$-T_r/T_s$	$1/(1 + T_s/T_r)$
Torque factor	$-T_s/T_r$	$1 + T_s/T_r$

b. Torque Factors. From the relationships in § 2, the torque factors are

$$TF_1 = Tq_{\text{out 1}}/Tq_{\text{in}} = -T_s/T_r, \text{ and}$$
$$TF_2 = Tq_{\text{out 2}}/Tq_{\text{in}} = 1 + T_s/T_r$$

which are the values of the TF in Conditions 2 and 4 of the simple systems and the reciprocals of the values in Condition 1, above.

From these, $Tq_{\text{out 2}}/Tq_{\text{out 1}}$ is a constant for any particular gear set and is determined by the ratio T_s/T_r. The value is the same as that of U in Condition 1, or $-(1 + T_r/T_s)$.

c. Speed Factors. The work equation (Chapter 1, § 3), divided by $(2\pi)(\text{Time})$, is

$$Tq_{\text{in}} S_{\text{in}} = (-Tq_{\text{out 1}})(-S_{\text{out 1}}) + (Tq_{\text{out 2}})(S_{\text{out 2}})$$

Dividing this equation by Tq_{in} and substituting the TF_1 and TF_2 values above for $Tq_{\text{out 1}}/Tq_{\text{in}}$ and $Tq_{\text{out 2}}/Tq_{\text{in}}$,

$$S_{\text{in}} = (T_s/T_r)(-S_{\text{out 1}}) + (1 + T_s/T_r)(S_{\text{out 2}})$$

Let $B = S_{\text{out 2}}/S_{\text{out 1}}$, where B may be zero or any negative number since the rotations of the output shafts always are in opposite directions. Then,

$$S_{\text{in}} = (T_s/T_r)(-S_{\text{out 1}}) + (1 + T_s/T_r)(B\,S_{\text{out 1}}), \text{ or}$$
$$SF_1 = S_{\text{out 1}}/S_{\text{in}} = 1/([T_s/T_r][B - 1] + B)$$

for a specified value of B.

From Chapter 1, § 2,

$$SF_2 = B\,SF_1, \text{ so}$$
$$SF_2 = 1/([T_s/T_r][1 - 1/B] + 1)$$

where B has the value above.

The numerical value of B is determined partially by the relationships between the values of the torques required by the loads connected to the output shafts and the speeds of the output shafts. The torque required by a load frequently increases with increases in the speed of rotation of the input shaft of the load, which is the same as the speed of the corresponding output shaft of the planetary gear system. Figure 14 illustrates the relationships between the output shaft speeds and torques for a particular gear system and pair of loads.

As stated above, Condition 2 is the inverse of Condition 1. If B in Condition 2 equals A in Condition 1, then SF_1 and SF_2 in Condition 2 equal $1/SF_1$ and $1/SF_2$ in Condition 1.

73

5. Conditions 3 and 5

a. SHAFT CONDITIONS (Fig. 11):

Shaft	Condition 3	Condition 5
Sun gear	Output	Input 1
Planet carrier	Input 1	Output
Ring gear	Input 2	Input 2

Condition 3 is a combination of Conditions 2 and 6 of the simple systems in Chart 1 and Condition 5 is a combination of Conditions 4 and 5 in that chart.

The determinations of the speed and torque factors in Conditions 3 and 5 of the compound systems are made in the same manner as in Condition 1 in § 3. When the computations corresponding with those in § 3 are performed, the results are as stated in the following.

b. SPEED FACTORS. The speed factors in Condition 3 are:

$$SF_1 = S_{out}/S_{in\,1} = (T_r/T_s)\,(1-A) + 1, \text{ and}$$
$$SF_2 = S_{out}/S_{in\,2} = (T_r/T_s)\,(1/A - 1) + 1/A,$$

for a specified value of A where $A = \pm S_{in\,2}/S_{in\,1}$, in which the $+$ sign indicates that input shaft 2 rotates in the forward direction (the direction of input shaft 1) and the $-$ sign indicates rotation in the reverse direction.

The speed factors in Condition 5 are:

$$SF_1 = S_{out}/S_{in\,1} = (T_s + A\,T_r)/(T_s + T_r), \text{ and}$$
$$SF_2 = S_{out}/S_{in\,2} = (T_s/A + T_r)/(T_s + T_r)$$

for a specified value of A where A has the value above.

For the reasons indicated in § 3, A must be negative in Condition 3 and positive in Condition 5. However, the above speed factors apply whether A has these signs or not.

c. TORQUE FACTORS. The torque factors in Condition 3 are:

$$TF_1 = Tq_{out}/Tq_{in\,1} = 1/(1 + T_r/T_s), \text{ and}$$
$$TF_2 = Tq_{out}/Tq_{in\,2} = -T_s/T_r, \text{ and}$$
$$U = \pm Tq_{in\,2}/Tq_{in\,1} = TF_1/TF_2 = -1/(1 + T_s/T_r)$$

The torque factors in Condition 5 are:

$$TF_1 = Tq_{out}/Tq_{in\,1} = 1 + T_r/T_s, \text{ and}$$
$$TF_2 = Tq_{out}/Tq_{in\,2} = 1 + T_s/T_r, \text{ and}$$
$$U = \pm Tq_{in\,2}/Tq_{in\,1} = TF_1/TF_2 = T_r/T_s$$

As in Condition 1, some of the above values apply only as long as Condition 3 or Condition 5 exists. The discussion in § 3 indicates the factors which cause these conditions not to exist.

6. Conditions 4 and 6

a. SHAFT CONDITIONS (Fig. 11):

Shaft	Condition 4	Condition 6
Sun gear	Input	Output 1
Planet carrier	Output 1	Input
Ring gear	Output 2	Output 2

Condition 4 is a combination of Conditions 1 and 5 of the simple systems in Chart 1 and Condition 6 is a combination of Conditions 3 and 6 in that chart.

The determinations of the speed and torque factors in Conditions 4 and 6 of the compound systems are made in the same manner as in Condition 2 in § 4. When the computations corresponding with those in § 4 are performed, the results are as stated in the following.

b. SPEED FACTORS. The speed factors in Condition 4 are:

$$SF_1 = S_{out\ 1}/S_{in} = 1/([T_r/T_s][1-B]+1), \text{ and}$$
$$SF_2 = S_{out\ 2}/S_{in} = 1/([T_r/T_s][1/B-1]+1/B)$$

for a specified value of B where $B = S_{out\ 2}/S_{out\ 1}$ and is always zero or a negative number. The numerical value of B is determined by the relative values of the various torques as discussed in § 4 and illustrated in Fig. 14.

The speed factors in Condition 6 are:

$$SF_1 = S_{out\ 1}/S_{in} = (T_s + T_r)/(T_s + B\,T_r), \text{ and}$$
$$SF_2 = S_{out\ 2}/S_{in} = (T_s + T_r)/(T_s/B + T_r)$$

for a specified value of B where B is always zero or a positive number. The numerical value of B is determined as in the case of Condition 4.

Conditions 4 and 6 are the inverse of Conditions 3 and 5, respectively. If B in Condition 4 equals A in Condition 3, SF_1 and SF_2 in Condition 4 equal $1/SF_1$ and $1/SF_2$ in Condition 3. Likewise, if B in Condition 6 equals A in Condition 5, SF_1 and SF_2 in Condition 6 equal $1/SF_1$ and $1/SF_2$ in Condition 5.

c. TORQUE FACTORS. The torque factors in Condition 4 are:

$$TF_1 = Tq_{out\ 1}/Tq_{in} = 1 + T_r/T_s, \text{ and}$$
$$TF_2 = Tq_{out\ 2}/Tq_{in} = -T_r/T_s$$

75

The torque factors in Condition 6 are:

$$TF_1 = Tq_{out\ 1}/Tq_{in} = 1/(1 + T_r/T_s), \text{ and}$$
$$TF_2 = Tq_{out\ 2}/Tq_{in} = 1/(1 + T_s/T_r)$$

7. Values of A and U

a. In this subparagraph, § 7, a, it is assumed that (1) Each of the input shafts in Condition 1, 3, or 5 has a power source which is independent of the other power source, (2) each power source provides sufficient torque at all times to prevent its being driven in the direction opposite to its normal direction of rotation, which is the direction needed to provide the proper sign of A in Condition 1, 3, or 5, and (3) the load at no time tends to drive the output shaft.

Assumptions (2) and (3) are used in order that Condition 1, 3, or 5 will exist rather than Condition 2, 4, or 6.

Then the work equation (Chapter 1, § 3), divided by (2π) (Time), is

$$S_{in\ 1}\ Tq_{in\ 1} + (\pm S_{in\ 2})\ (\pm Tq_{in\ 2}) = S_{out}\ Tq_{out}$$

Since

$$A = \pm S_{in\ 2}/S_{in\ 1} \quad \text{and} \quad U = \pm Tq_{in\ 2}/Tq_{in\ 1}$$
$$S_{in\ 1}\ Tq_{in\ 1} + A\ S_{in\ 1}\ U\ Tq_{in\ 1} = S_{out}\ Tq_{out}$$

where A and U are negative in Conditions 1 and 3 and positive in Condition 5.

Dividing this equation by $S_{in\ 1}\ Tq_{in\ 1}$,

$$1 + A\ U = SF_1\ TF_1, \text{ or}$$
$$A\ U = SF_1\ TF_1 - 1$$

Again substituting in the work equation but in terms of $S_{in\ 2}$, etc., and dividing by $S_{in\ 2}\ Tq_{in\ 2}$, the result is

$$1/(A\ U) = SF_2\ TF_2 - 1$$

The above expressions may be used in determining the correctness of the values of SF_1, TF_2, U, etc., in each of Conditions 1, 3, and 5. When these are correct, use of the expressions results in equations such as

$$A\ T_s/T_r = A\ T_s/T_r, \text{ or}$$
$$1 = 1$$

b. Figure 15 illustrates the relationships between the torques and the speeds of a compound system in Condition 5. In this figure, power source 1 has no controls other than an on-off control, but source 2 has a control by which its output torque-vs.-speed characteristic may be changed. These characteristics for three adjustments of the control are illustrated in the figure as adjustments 1, 2, and 3.

When adjustment 2 is used, the output torques of sources 1 and 2 have maximum values at power source 1 and 2 speeds of 700 and 175 rpm, respectively. The value of $\pm Tq_{in\,2}/Tq_{in\,1}$ is $20/10 = 2$, which is the value of U in Condition 5 when $T_r = 40$ and $T_s = 20$. The value of A, then, is 0.25 and S_{out} is 350 rpm.

If the control of source 2 is changed to adjustment 1, the maximum torque of this source is 15 lb-ft and it occurs when $S_{in\,2}$ is 150 rpm. If the speeds of sources 1 and 2 are 700 and 150 rpm,

$$S_{out} = (700)\,(1/3) + (150)\,(2/3) = 333 \text{ rpm}$$

This output speed requires output torques of 18 and 9 lb-ft from sources 2 and 1. Source 2 cannot provide the torque required from it and source 1 can provide a greater torque than is required. The ratio, $Tq_{in\,2}/Tq_{in\,1}$, is 15/10 or 1.5, which is less than it must be in Condition 5. Therefore, power source 1 would place a greater torque on input shaft 2 than power source 2 can provide, and source 1 would tend to drive source 2 in the reverse direction. This would reduce the speed of source 2, which in turn would reduce its torque capability. The reduction in speed would tend to cause a reduction in S_{out}, and the above excess torque capability of source 1 would increase. $S_{in\,1}$, then, increases, which tends to increase S_{out}, and the deficit in $Tq_{in\,2}$ would tend to increase as a result of this. Let it be assumed that the maximum speed of source 1 is 1,000 rpm and that $S_{in\,1}$ increases to this value. The power capability of source 1, then, is

$$P = Tq\,S\,2\pi = 6{,}000\,(2\pi) \text{ ft-lb per minute}$$

If all of this power were transferred to the output shaft, the speed and the torque of this shaft would be approximately 293 rpm and 20.5 lb-ft. However, the output torque of source 1 when $S_{in\,1}$ is 1,000 rpm is 6 lb-ft, and this torque produces only 18 lb-ft of torque on the output shaft of the system. Therefore, one or both of two events must occur: (1) the speed of source 1 may decrease and thereby increase the source 1 torque capability, (2) the speed of the output shaft may decrease and thereby decrease the torque requirement placed on source 1. Let it be assumed that the output shaft speed decreases to 280 rpm, which requires 6 lb-ft of torque from source 1, and that $S_{in\,1}$ remains 1,000 rpm. The equation of the speeds would be

$$S_{out} = (S_{in\,1})\,(\tfrac{1}{3}) + (S_{in\,2})\,(\tfrac{2}{3}), \text{ or}$$
$$S_{in\,2} = (280 - 333)\,(\tfrac{3}{2}) = -80 \text{ rpm}$$

Since the torque placed on the source 2 shaft by source 1 is in the reverse direction and the rotation of source 2 is in the reverse direction, power source 2 acts as a load on the system rather than as a source of power. Some of the power from source 1 is used in driving source 2 and not all of it drives the load on the planet carrier as is assumed in the

77

above. The result of the conditions above is that power source 1 drives the load and power source 2, and the ratio of the speeds of the latter two is determined by their torque-*vs.*-speed characteristics. These characteristics in the case of source 2 are those which apply to it when it is driven, rather than those shown in Figure 15. Condition 4 in Chart 3 exists rather than Condition 5.

If the control is moved to the adjustment 3 position, the ratio of the maximum torque capabilities of the sources is

$$Tq_{in\ 2}/Tq_{in\ 1} = 2.4$$

Power source 2, then, would drive power source 1 and the latter would act as a load on the system rather than as a source of power, and Condition 2 in Chart 3 would exist.

Figure 16 illustrates the relationships between S_{out}, $S_{in\ 1}$, and A when the torque required by the load is such that the torques required from the power sources are within the torque capabilities of the sources at all of the speeds shown in the chart. In this chart, the speed of one or each of the power sources is controlled to provide the desired values of A and S_{out}.

Figures 15 and 16 illustrate Condition 5, but the principles involved are applicable also to Conditions 1 and 3.

c. As stated before, A and U must be negative in Conditions 1 and 3 and positive in Condition 5. If such is not the case, the TF_1, TF_2, and U values stated in the foregoing are not correct in some instances. However, the speed factors are correct regardless of the signs of A and U.

In some instances, the speed ranges of the available power sources may be limited to certain high values and a low output shaft speed may be desired. In these cases, the system may be operated as though it were in Condition 1 or 3 but with positive values of A, or in Condition 5 but with negative values of A. Then, as shown by the use of the equation in § 3,

$$S_{out} = (S_{in\ 1})\ (SF\ of\ corresponding\ simple\ system)\ +$$
$$(\pm S_{in\ 2})\ (SF\ of\ corresponding\ simple\ system)$$

The speed of the output shaft may be made to be less than that of either of the input shaft speeds, and the direction of rotation of the output shaft may be changed from forward to reverse and vice versa by changing one or the other (or both) of the power source speeds. In operations such as these, one of the power sources is driven to varying extents by the other. This, of course, means that the power efficiency of the system is low, particularly when prime movers are used as power sources.

A variation of the above may be used with only one power source. It may be connected to the gear system so as to provide a value of A

which results in the desired S_{out}/S_{source} ratio. As in the preceding cases, the power source of one input shaft tends to drive the power source of the other, but, since only one source is used, the only power loss is that due to friction in the gear set and in such other devices as are used in the power train between the power source and the gear set.

d. Figures 26 and 27 illustrate a transmission in which Condition 5 is used when in High gear. Both input shafts are driven by the same power source, the engine. The torque converter provides the means by which U is maintained at the value corresponding with Condition 5, which is $T_r/T_s = 1.74$. The following indicates the manner in which this value is maintained. (Throttle opening and road conditions are assumed to be constants.)

The load torque is equal in magnitude to the torque required to drive the automobile at its speed and to provide such acceleration of the automobile as exists, but it has the opposite sign. The output torque of the gear set when the automobile speeds are constants is of a form resembling

$$Tq_{out} = C \, S^x$$

where C is a constant, S is the output shaft speed, and x is a value which is greater than 1.0. For the purposes of this discussion, x is assumed to be 2.0 within the very small range of speeds of the output shaft involved in the discussion. Thus, a change in speed of the output shaft of a few per cent requires a larger percentage change in the output torque when the speeds before and after the change are constant speeds. During a reduction or increase in speed, the inertia of the automobile tends to maintain the output shaft speed constant. Therefore, less or more torque is required from the engine and gear set during a slow deceleration or acceleration of the automobile than in constant-speed operation.

Let it be assumed that the ratio of the torques applied to the ring gear and the sun gear shown in Fig. 27 becomes less than the value of U, or 1.74. Then the torque on the sun gear shaft would tend to drive the turbine in the reverse direction. As a result, the speed of the turbine would decrease, its slip would increase, and the torque produced by the turbine on the ring gear would increase. Thus, the ratio of the torques applied to the ring and the sun gears would increase to the value of U.

Now assume that the ratio of the torques becomes greater than the value of U. Then the torque on the ring gear would tend to rotate the sun gear in the reverse direction, and its speed, which is the speed of the engine, would tend to decrease. Likewise, the speed of the turbine, which provides the torque on the ring gear, would tend to increase. As a result, the slip of the turbine would decrease and the torque produced by the turbine would decrease. The ratio of the torques on the ring and the sun gears, then, would decrease to the value of U.

79

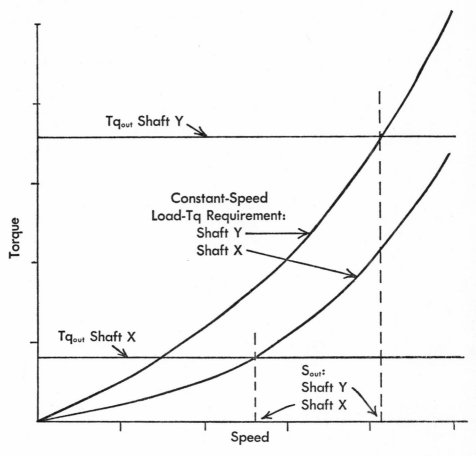

Fig. 14. Compound system output-shaft speeds-*vs.*-torque requirements. (Simple gear set—Fig. 11.) (1) Input shaft torque constant. (2) $S_{out\ 2}/S_{out\ 1} = B$ in Chart 3. (3) Directions of rotation and torques are disregarded. (4) Principles illustrated are applicable to Conditions 2, 4, and 6.

The above adjustments of $Tq_{in\ 2}/Tq_{in\ 1}$ occur in a continuous manner, and this ratio is maintained at its value in Condition 5 of 1.74.

Figure 33 illustrates another transmission in which two gear sets are in Condition 5 when in 4th gear. The torque adjustment devices in these cases are fluid clutches.

8. Summary and Graphs

The results of the computations of speed and torque factors are included in Chart 3. The torque factors are shown in graphic form in Figs. 12 and 13. Chart 2 indicates the applicability of the two figures. The relationships between the speeds of the output shafts in Conditions

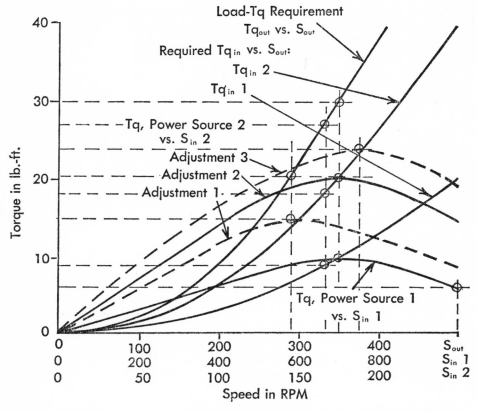

FIG. 15. Compound system Condition 5 speed-*vs.*-torque relationships. (Simple gear set—Fig. 11.) (1) $T_r = 40$, $T_s = 20$. (2) Principles illustrated are applicable to Conditions 1, 3, and 5.

2, 4, and 6 and the various torque values are illustrated in Fig. 14. The corresponding relationships in Conditions 1, 3, and 5 are illustrated in Figs. 15 and 16.

CHART 3. COMPOUND PLANETARY GEAR SYSTEMS
(Simple Gear Set—Fig. 11)

| Cond. | Shaft | | | SF_1 | TF_1 | | |
	S	PC	R		Value	Fig.	Form
1	I_1	I_2	O	$(T_s/T_r)(A-1)+A$ *	$-T_r/T_s$	12	1
2	O_1	O_2	I	$1/((T_s/T_r)(B-1)+B)$ *	$-T_s/T_r$	12	1
3	O	I_1	I_2	$(T_r/T_s)(1-A)+1$ *	$1/(1+T_r/T_s)$	13	3
4	I	O_1	O_2	$1/((T_r/T_s)(1-B)+1)$ *	$1+T_r/T_s$	13	2
5	I_1	O	I_2	$(T_s+AT_r)/(T_s+T_r)$ **	$1+T_r/T_s$	13	2
6	O_1	I	O_2	$(T_s+T_r)/(T_s+BT_r)$ **	$1/(1+T_r/T_s)$	13	3

81

CHART 3. COMPOUND PLANETARY GEAR SYSTEMS *(cont.)*

				SF_2	TF_2		
1	I_1	I_2	O	$(T_s/T_r)(1-1/A)+1$ *	$1/(1+T_s/T_r)$	13	3
2	O_1	O_2	I	$1/((T_s/T_r)(1-1/B)+1)$ *	$1/(1+T_s/T_r)$	13	2
3	O	I_1	I_2	$(T_r/T_s)(1/A-1)+1/A$ *	$1+T_s/T_r$	12	1
4	I	O_1	O_2	$1/((T_r/T_s)(1/B-1)+1/B)$ *	$-T_s/T_r$	12	1
5	I_1	O	I_2	$(T_s/A+T_r)/(T_s+T_r)$ **	$-T_r/T_s$	13	2
6	O_1	I	O_2	$(T_s+T_r)/(T_s/B+T_r)$ **	$1+T_s/T_r$	13	3

Notes: $SF_1 = S_{out}/S_{in\,1}$ or $S_{out\,1}/S_{in}$ * A or B always negative
        $SF_2 = S_{out}/S_{in\,2}$ or $S_{out\,2}/S_{in}$ ** A or B always positive
        $TF_1 = Tq_{out}/Tq_{in\,1}$ or $Tq_{out\,1}/Tq_{in}$
        $TF_2 = Tq_{out}/Tq_{in\,2}$ or $Tq_{out\,2}/Tq_{in}$
        $A = S_{in\,2}/S_{in\,1}$; $B = S_{out\,2}/S_{out\,1}$

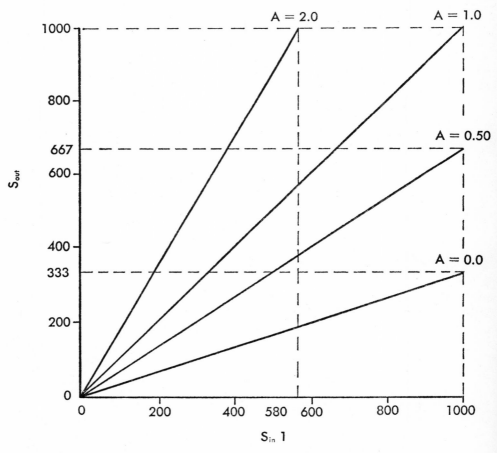

FIG. 16. Compound system Condition 5 speed relationships. (Simple gear set—Fig. 11.) (1) $T_r = 40$, $T_s = 20$. (2) Required Tq_{in} values are less than Tq capabilities of power sources. (3) Principles illustrated are applicable to Conditions 1, 3, and 5.

Simple Planetary Gear Systems Employing Compound Gear Set With Ring Gear

1. Conditions of Operation

There are 60 possible Conditions of Operation of simple planetary gear systems which employ a compound gear set of the type shown in Fig. 17, or of the type indicated by the note in that figure. The shafts used as an input shaft, etc., in each of the 60 conditions are listed in Chart 4, where *I, O, H*, and *F* mean Input, Output, Held (stationary), and Free (to rotate).

2. Conditions 1–24

    a. Shaft Conditions (Fig. 17):

One shaft	—	Input
Another shaft	—	Output
Third shaft	—	Held
Fourth shaft	—	Free

    b. Conditions with Sun Gear 2 Free. These conditions are the same as Conditions 1–6 of the simple systems which employ the simple gear set shown in Fig. 11 since sun gear 1 in Fig. 17 is the equivalent of the sun gear shown in Fig. 11. They are entered in Chart 4 as Conditions 1–6. The *SF* and *TF* values are the same as in Conditions 1–6 in Chart 1. However, they are recomputed in the following in the process of computing the factors pertaining to other conditions.

    c. Conditions with Sun Gear 2 Held. Assume that the planet carrier turns one revolution in the forward direction. Then pinions 2 make T_{s2}/T_{p2} revolutions in the forward direction and pinions 1 make

$$(T_{s2}/T_{p2})\,(-T_{p2}/T_{p1}) = -T_{s2}/T_{p1} \text{ revolutions.}$$

83

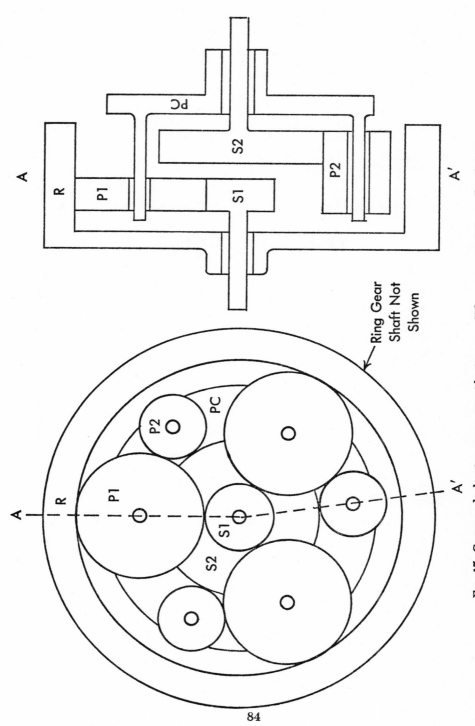

Fig. 17. Compound planetary gear set with ring gear. When sun gear 2 is smaller than sun gear 1, *P*1 are long pinions and *P*2 are short pinions.

As a result of the pinion rotations, sun gear 1 turns

$$(-T_{s2}/T_{p1}) \, (-T_{p1}/T_{s1}) = T_{s2}/T_{s1} \text{ revolutions}$$

In addition, sun gear 1 would make one revolution in the forward direction as the result of the revolution of the planet carrier if the pinions did not rotate, but pinions 2 slid along the surface of sun gear 2. The total number of revolutions of sun gear 1 that result from one revolution of the planet carrier is the sum of the two components of the rotation, or

$$1 + T_{s2}/T_{s1}$$

The revolution of the planet carrier would cause one revolution in the forward direction of the ring gear if the pinions did not rotate. In addition, the ring gear turns

$$(-T_{s2}/T_{p1}) \, (T_{p1}/T_r) = -T_{s2}/T_r \text{ revolutions}$$

as the result of the pinion rotations. The total number of revolutions of the ring gear that results from one revolution of the planet carrier is the sum of the two components, or

$$1 - T_{s2}/T_r$$

Since all of the above revolutions occur in the same time period, the ratios of the speeds of the rotating elements are:

$$S_{s1}/S_{pc} = (1 + T_{s2}/T_{s1})/1 = 1 + T_{s2}/T_{s1},$$
$$S_r/S_{pc} = (1 - T_{s2}/T_r)/1 = 1 - T_{s2}/T_r,$$

and dividing the first ratio by the second,

$$S_{s1}/S_r = (1 + T_{s2}/T_{s1})/(1 - T_{s2}/T_r)$$

The first ratio is the SF of the system when the planet carrier shaft is the input shaft, the shaft of sun gear 1 is the output shaft, and the ring gear shaft is Free. The second is the SF when the planet carrier shaft is the input shaft, the ring gear shaft is the output shaft, and the sun gear 1 shaft is Free. Likewise, the third ratio is the SF when the input, output, and Free shafts are those of the ring gear, sun gear 1, and planet carrier. In all of these conditions, sun gear 2 is Held.

The reciprocals of the above ratios are the SF values when the inverse conditions exist; that is, when the functions of the shafts serving as input and output shafts in the above are exchanged.

The torque factor in each of the conditions is the reciprocal of the speed factor since the systems are simple systems (Chapter 1, § 1).

The above conditions and the inverse conditions are entered in Chart 4 as Conditions 13–18.

d. CONDITIONS WITH PLANET CARRIER HELD. Assume that sun gear 1 turns one revolution in the forward direction. Then pinions 1 make $-T_{s1}/T_{p1}$ revolutions and pinions 2 make

$$(-T_{s1}/T_{p1})\,(-T_{p1}/T_{p2}) = T_{s1}/T_{p2} \text{ revolutions}$$

The ring gear turns

$$(-T_{s1}/T_{p1})\,(T_{p1}/T_r) = -T_{s1}/T_r \text{ revolutions}$$

and sun gear 2 turns

$$(T_{s1}/T_{p2})\,(-T_{p2}/T_{s2}) = -T_{s1}/T_{s2} \text{ revolutions}$$

The ratios of the speeds of the rotating elements are:

$$S_r/S_{s1} = -T_{s1}/T_r,$$
$$S_{s2}/S_{s1} = -T_{s1}/T_{s2}, \text{ and}$$
$$S_r/S_{s2} = (-T_{s1}/T_r)/(-T_{s1}/T_{s2}) = T_{s2}/T_r$$

These ratios are the speed factors of the systems when the output and input shafts are: the ring and sun gear 1 shafts, the shafts of sun gears 2 and 1, the ring and sun gear 2 shafts. In each condition the planet carrier is Held and the fourth shaft is Free.

The speed factors in the inverse conditions are the reciprocals of the above ratios, and the torque factor in each condition is the reciprocal of the speed factor.

The above conditions and the inverse conditions are entered in Chart 4 as Conditions 1, 2, 7, 8, 23, and 24.

e. CONDITIONS WITH RING GEAR HELD. Assume that the planet carrier turns one revolution in the forward direction. Then pinions 1 make $-T_r/T_{p1}$ revolutions and pinions 2 make

$$(-T_r/T_{p1})(-T_{p1}/T_{p2}) = T_r/T_{p2} \text{ revolutions}$$

Sun gear 1 makes

$$(-T_r/T_{p1})(-T_{p1}/T_{s1}) = T_r/T_{s1} \text{ revolutions}$$

and sun gear 2 makes

$$(T_r/T_{p2})(-T_{p2}/T_{s2}) = -T_r/T_{s2} \text{ revolutions}$$

as the results of the rotations of pinions 1 and 2.

In addition, each of the sun gears would make one revolution in the forward direction if the pinions did not rotate, but pinions 1 slid along the surface of the ring gear.

The total numbers of revolutions made by the sun gears are the sums of the components of rotation, or: sun gear 1, $1 + T_r/T_{s1}$; sun gear 2, $1 - T_r/T_{s2}$.

The ratios of the speeds of the rotating elements are:

$$S_{s1}/S_{pc} = 1 + T_r/T_{s1},$$
$$S_{s2}/S_{pc} = 1 - T_r/T_{s2}, \text{ and}$$
$$S_{s1}/S_{s2} = (1 + T_r/T_{s1})/(1 - T_r/T_{s2})$$

As in the preceding cases, these are the speed factors in three of the conditions of operation, and the torque factors in these conditions are the reciprocals of the speed factors.

The above conditions and the inverse conditions are entered in Chart 4 as Conditions 5, 6, 19, 20, 21, and 22.

f. Conditions with Sun Gear 1 Held. Assume that the planet carrier turns one revolution in the forward direction. Then pinions 1 make T_{s1}/T_{p1} revolutions in the forward direction and pinions 2 make

$$(T_{s1}/T_{p1})(-T_{p1}/T_{p2}) = -T_{s1}/T_{p2} \text{ revolutions}$$

The ring gear makes

$$(T_{s1}/T_{p1})(T_{p1}/T_r) = T_{s1}/T_r \text{ revolutions}$$

and sun gear 2 makes

$$(-T_{s1}/T_{p2})(-T_{p2}/T_{s2}) = T_{s1}/T_{s2} \text{ revolutions}$$

as the results of the rotations of pinions 1 and 2.

The ring and sun 2 gears also would turn one revolution each in the forward direction as the result of the planet carrier revolution if the pinions did not rotate but pinions 1 slid along the surface of sun gear 1.

The total numbers of revolutions of the ring and sun 2 gears are: ring gear, $1 + T_{s1}/T_r$; sun gear 2, $1 + T_{s1}/T_{s2}$.

The ratios of the speeds of the rotating elements are:

$$S_r/S_{pc} = 1 + T_{s1}/T_r,$$
$$S_{s2}/S_{pc} = 1 + T_{s1}/T_{2s}, \text{ and}$$
$$S_r/S_{s2} = (1 + T_{s1}/T_r)/(1 + T_{s1}/T_{s2})$$

The conditions of operation pertaining to these ratios and the inverse conditions are entered in Chart 4 as Conditions 3, 4, 9, 10, 11, and 12.

3. Conditions 25–36

a. Shaft Conditions (Fig. 17):

Any shaft   —   Input
Any other shaft   —   Output
Remaining shafts   —   Free

b. Speed and Torque Factors. Analysis of the system in each of the above conditions shows that the only cause for rotation of the output shaft is the friction within the system. Assuming an appreciable load on the output shaft and negligible friction within the system, the output shaft speed is zero. Then, in each condition,

$$SF = 0.0, \text{ and } TF = \text{(substantially) } 0.0$$

4. Conditions 37–48

a. SHAFT CONDITIONS (Fig. 17):

Any shaft	—	Input
Any other shaft	—	Output
Any two or more shafts	—	Held

b. SPEED AND TORQUE FACTORS. Analysis of the system in each of the above conditions shows that the input shaft cannot rotate. Therefore, the output shaft does not rotate and has no torque that can be employed in driving a load. Then, the SF and TF may be considered to be zero.

5. Conditions 49–60

a. SHAFT CONDITIONS (Fig. 17):

Any shaft	—	Input
Any other shaft	—	Output
Any two or more shafts	—	Locked together

b. SPEED AND TORQUE FACTORS. In Conditions 37–48, the two or more shafts are locked together indirectly since each is locked to a stationary object. As a result, the input and output shafts cannot rotate.

In Conditions 49–60, a locked situation exists but no stationary object is involved. As a result, the entire gear set rotates with the input shaft, and the output shaft rotates with the gear set. Therefore,

$$SF = 1.0 \text{ and } TF = 1.0$$

6. Summary and Graphs

The speed and torque factors in Conditions 1–60 are indicated in Chart 4. Also in that chart are references to figures which have graphs of the speed factors, or speed and torque factors in the case of Fig. 18. Since $TF = 1/SF$ in simple systems, the references to Fig. 12, etc., and Form 1, etc., may be used in determining which one of Figs. 12 and 13, and which Form, is a graph of a particular torque factor. For example, Form 2 of Fig. 13 is the reference for the Condition 3 speed factor graph. The torque factor in Condition 3 is $1/SF$ in Condition 3, or $1/(1 + T_{s1}/T_r)$. Form 3 in Fig. 13 is the reference for a graph of this quantity, so it may be employed as a graph of the torque factor in Condition 3.

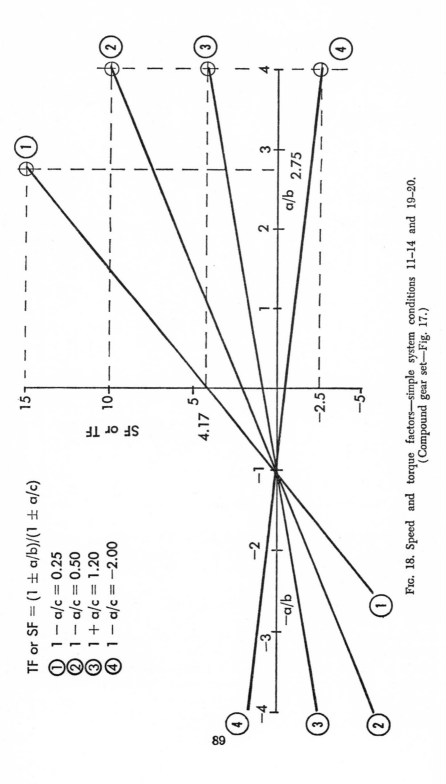

TF or SF = $(1 \pm a/b)/(1 \pm a/c)$

① $1 - a/c = 0.25$
② $1 - a/c = 0.50$
③ $1 + a/c = 1.20$
④ $1 - a/c = -2.00$

SF or TF

Fig. 18. Speed and torque factors—simple system conditions 11–14 and 19–20.
(Compound gear set—Fig. 17.)

89

CHART 4. SIMPLE PLANETARY GEAR SYSTEMS
(Compound Gear Set—Fig. 17)

	Shaft					Graph	
Cond.	S_1	PC	R	S_2	SF	Chart	Form
1	I	H	O	F	$-T_{s1}/T_r$	12	1
2	O	H	I	F	$-T_r/T_{s1}$	12	1
3	H	I	O	F	$1+T_{s1}/T_r$	13	2
4	H	O	I	F	$1/(1+T_{s1}/T_r)$	13	3
5	I	O	H	F	$1/(1+T_r/T_{s1})$	13	3
6	O	I	H	F	$1+T_r/T_{s1}$	13	2
7	I	H	F	O	$-T_{s1}/T_{s2}$	12	1
8	O	H	F	I	$-T_{s2}/T_{s1}$	12	1
9	H	I	F	O	$1+T_{s1}/T_{s2}$	13	2
10	H	O	F	I	$1/(1+T_{s1}/T_{s2})$	13	3
11	H	F	I	O	$(1+T_{s1}/T_{s2})/(1+T_{s1}/T_r)$	18	–
12	H	F	O	I	$(1+T_{s1}/T_r)/(1+T_{s1}/T_{s2})$	18	–
13	I	F	O	H	$(1-T_{s2}/T_r)/(1+T_{s2}/T_{s1})$	18	–
14	O	F	I	H	$(1+T_{s2}/T_{s1})/(1-T_{s2}/T_r)$	13	–
15	F	I	O	H	$1-T_{s2}/T_r$	13	2
16	F	O	I	H	$1/(1-T_{s2}/T_r)$	18	3
17	I	O	F	H	$1/(1+T_{s2}/T_{s1})$	13	3
18	O	I	F	H	$1+T_{s2}/T_{s1}$	13	2
19	I	F	H	O	$(1-T_r/T_{s2})/(1+T_r/T_{s1})$	18	–
20	O	F	H	I	$(1+T_r/T_{s1})/(1-T_r/T_{s2})$	18	–
21	F	I	H	O	$1-T_r/T_{s2}$	13	2
22	F	O	H	I	$1/(1-T_r/T_{s2})$	13	3
23	F	H	I	O	T_r/T_{s2}	12	1
24	F	H	O	I	T_{s2}/T_r	12	1
25–36	Two Free				$SF=0.0 \quad TF=0.0$	–	–
37–48	Two Held				$SF=0.0 \quad TF=0.0$	–	–
49–60	Two locked together				$SF=1.0 \quad TF=1.0$	–	–

Note: $TF=1/SF$ except in Conditions 25–48.

The principles discussed in Chapter 5, § 2, are applicable to the torques in Conditions 1–60 in this chapter.

In Appendix C is a discussion of a nonplanetary gear system which is the equivalent of the planetary gear set shown in Fig. 17.

7

Compound Planetary Gear Systems Employing Compound Gear Set With Ring Gear

1. Conditions of Operation

There are 62 possible conditions of operation of the gear set shown in Fig. 17 in compound systems, not including those in which two shafts are locked together (Chart 4).

2. Speed and Torque Factors

The usual employment of planetary gear sets of the type shown in Fig. 17 is in simple systems. For this reason, computations of the speed and torque factors in compound systems are omitted. However, comments are made in the following concerning these factors.

The values of the factors in some of the conditions are obvious. In others, restrictions pertaining to the relative speeds of the shafts are obvious.

a. CONDITIONS 1–6 (SUN GEAR 2 FREE). These conditions are the equivalents of Conditions 1–6 in Chapter 5 pertaining to the use of the gear set shown in Fig. 11 in compound systems since sun gear 2, being Free, does not affect the operation of the gear set. Therefore, the speed and torque factors are as stated in Chart 3.

b. CONDITIONS 7–30 (ONE SHAFT HELD). The speed factors in these conditions may be obtained from the computations in Chapter 6, § 2. As an example, assume that the shaft conditions in a compound system are:

Sun gear 1 shaft   —   Output 1
Planet carrier shaft   —   Output 2

Ring gear shaft   —   Input
Sun gear 2 shaft   —   Held

From § 2, c, of Chapter 6,

$$S_{s1}/S_r = SF_1 = (1 + T_{s2}/T_{s1})/(1 - T_{s2}/T_r), \text{ and}$$
$$S_{pc}/S_r = SF_2 = 1/(1 - T_{s2}/T_r)$$

It should be noted that the ratio of the speeds of the output shafts in this instance is

$$S_{\text{out } 2}/S_{\text{out } 1} = SF_2/SF_1 = 1/(1 + T_{s2}/T_{s1})$$

which is a constant for any particular gear set. Therefore, this ratio, which corresponds with the value B in Chapter 5, is not a function of the torques presented by the loads on output shafts 2 and 1, as in the case of compound systems which employ the gear set shown in Fig. 11.

From the above, the relationships between the input and output shaft speeds, and between the latter two speeds, are determined solely by the ratios of the numbers of teeth on the gears. Then the torque factors of the systems in Conditions 7–30 are not determined solely by the gear-teeth ratios as they are in Chapter 5. Instead, they also are functions of the torques the loads present to the output shafts. For example, using the above compound system, let it be assumed that the load connected to output shaft 1 presents negligible torque to the system and the torque presented by the load connected to output shaft 2 is not negligible. Then, substantially,

$$TF_2 = 1 - T_{s2}/T_r, \text{ and}$$
$$TF_1 = 0.0/Tq_{in} = 0.0$$

If the load-torque conditions are exchanged, the values are substantially

$$TF_1 = (1 - T_{s2}/T_r)/(1 + T_{s2}/T_{s1}), \text{ and}$$
$$TF_2 = 0.0/Tq_{in} = 0.0$$

c. CONDITIONS 31–62. Consider the compound system in which the sun gear 2 shaft is the output shaft and the other three shafts are input shafts. Chart 3 indicates the speed that any one of the input shafts in the compound system must have when the other two input shafts have particular speeds. For example, if the sun gear 1 and planet carrier input shafts have speeds of X and $-2X$, then the speed of the ring gear input shaft must have a value that is determined by the speed factor, SF_1, in Condition 1 in Chart 3, or

$$S_r = (X) \ ([T_{s1}/T_r] \ [-2-1] \ -2)$$

since A in that speed factor is $-2X/X = -2.0$. Therefore, the speeds of the input shafts cannot be independent of one another, but one must have a particular value when the other two have particular values.

Since the above restriction pertains to the speeds of the input shafts in this condition, it applies also to the speeds of the output shafts in the inverse condition for similar reasons.

The principles of the methods employed in preceding chapters are applicable to the computation of speed and torque factors in those instances in which the factors may not be obtainable by methods such as those indicated above.

Simple Planetary Gear Systems Employing Simple Gear Set Without Ring Gear

1. Numbers of Gear Teeth

Figure 19 shows that the sum of the radii of sun gear 1 and pinion 1 equals the sum of the radii of sun gear 2 and pinion 2. The number of teeth on each of the gears is proportional to the radius of the gear, assuming that all of the gears have the same number of teeth per foot of circumference. Then,

$$T_{s1} + T_{p1} = T_{s2} + T_{p2}$$

2. Conditions of Operation

There are 18 possible conditions of operation of simple planetary gear systems which employ the gear set shown in Fig. 19. These are listed in Chart 5.

3. Condition 1

a. SHAFT CONDITIONS (Fig. 19):

Sun gear 1 shaft	—	Output
Sun gear 2 shaft	—	Held
Planet carrier shaft	—	Input

b. SPEED FACTOR. When the input shaft turns one revolution, the planet pinions make T_{s2}/T_{p2} revolutions on their axes in the forward direction. If the pinions were stationary with respect to the input shaft axis when these revolutions are made, sun gear 1 would make T_{p1}/T_{s1} revolutions in the reverse direction during each pinion revolution. The resultant, with the pinions stationary, would be $(T_{s2}/T_{p2})(T_{p1}/T_{s1})$ revolutions of the output shaft in the reverse direction.

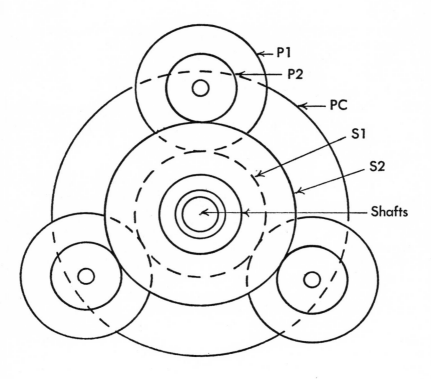

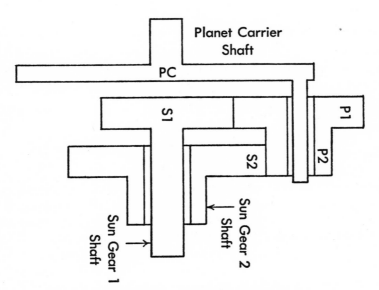

Fig. 19. Simple planetary gear set without ring gear.

95

However, the pinions do not remain stationary, but make one revolution about the axis of the input shaft with each revolution of that shaft. This revolution would cause one revolution of sun gear 1 if the planet pinions did not rotate about their axes, and this rotation would be in the forward direction.

The rotation of the output shaft due to both of the above components is $1 - (T_{s2} T_{p1})/(T_{s1} T_{p2})$ revolutions in the forward direction if this quantity is positive, and in the reverse direction if it is negative. The speed factor is

$$SF = S_{out}/S_{in} = 1 - (T_{s2} T_{p1})/(T_{s1} T_{p2})$$

In Fig. 19, T_{s2} is greater than T_{s1} and T_{p1} is greater than T_{p2}. The quantity involving these values, then, is greater than 1.0, and the direction of rotation of the output shaft is reverse. If the sun gears were of equal sizes, the planet pinions would be of equal sizes, the quantity would equal 1.0, and the SF would be zero. If sun gear 1 were larger than sun gear 2, pinion 2 would be larger than pinion 1, and the direction of rotation of the output shaft would be forward. The SF pertaining to the forward direction of rotation of the output shaft always is less than 1.0, while that pertaining to the reverse direction may have any value.

c. TORQUE FACTOR. The torque factor is

$$TF = Tq_{out}/Tq_{in} = 1/SF$$

4. Condition 2

a. SHAFT CONDITIONS (Fig. 19):

Sun gear 1 shaft	—	Input
Sun gear 2 shaft	—	Held
Planet carrier shaft	—	Output

b. SPEED AND TORQUE FACTORS. This condition is the inverse of Condition 1, so

$$TF = 1 - (T_{s2} T_{p1})/(T_{s1} T_{p2}), \text{ and}$$
$$SF = 1/TF$$

The direction of rotation of the output shaft may be forward or reverse, depending on the values of T_{s1}, etc. The SF pertaining to the forward direction always is greater than 1.0, while that pertaining to the reverse direction may have any value.

5. Conditions 3–4

a. SHAFT CONDITIONS (Fig. 19):

	Condition 3	Condition 4
Sun gear 1 shaft	Held	Held
Sun gear 2 shaft	Output	Input
Planet carrier shaft	Input	Output

b. CONDITION 3. This condition is similar to Condition 1. Applying the same process of computation,

$$SF = 1 - (T_{s1} T_{p2})/(T_{s2} T_{p1}), \text{ and}$$
$$TF = 1/SF$$

c. CONDITION 4. This condition is the inverse of Condition 3, so

$$TF = 1 - (T_{s1} T_{p2})/(T_{s2} T_{p1}), \text{ and}$$
$$SF = 1/TF$$

d. CONDITIONS 3 AND 4. The discussion in §§ 3 and 4 pertaining to the numerical values of the speed factors, with appropriate changes because of the differences in the speed factors, applies in Conditions 3 and 4.

6. Conditions 5–6

a. SHAFT CONDITIONS (Fig. 19):

	Condition 5	Condition 6
Sun gear 1 shaft	Input	Output
Sun gear 2 shaft	Output	Input
Planet carrier shaft	Held	Held

b. CONDITION 5. Each revolution of the input shaft causes the pinions to rotate T_{s1}/T_{p1} revolutions in the reverse direction. Each pinion revolution causes sun gear 2 and the output shaft to make T_{p2}/T_{s2} revolutions in the forward direction. Then,

$$SF = (T_{s1} T_{p2})/(T_{s2} T_{p1}), \text{ and}$$
$$TF = 1/SF$$

c. CONDITION 6. This condition is the inverse of Condition 5, so

$$SF = (T_{s2} T_{p1})/(T_{s1} T_{p2}), \text{ and}$$
$$TF = 1/SF$$

7. Conditions 7–12

a. SHAFT CONDITIONS (Fig. 19):

Any shaft	—	Input
Any other shaft	—	Output
Remaining shaft	—	Free

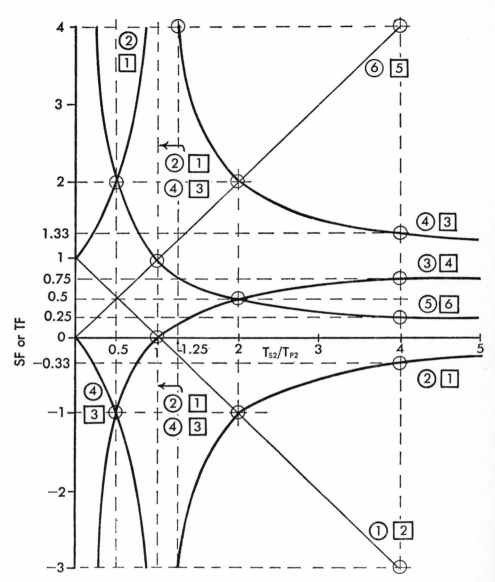

FIG. 20. Simple system speed and torque factors. (Simple gear set—Fig. 19.) (1) $(T_{p1}/T_{s1}) = 1.0$. (2)①to⑥–SF in Conditions 1–6. (3)[1] to[6]–TF in Conditions 1–6.

b. SPEED AND TORQUE FACTORS. These conditions are the same as Conditions 1–6 except that the Held shafts in those conditions are Free in Conditions 7–12. Reasoning similar to that pertaining to Conditions 7–12 in the case of the simple gear set shown in Fig. 11 in Chapter 4 shows

98

that the output shaft has (substantially) no torque, and zero speed if there is an appreciable load on the output shaft, in any of the above conditions. Therefore,

$$SF = 0.0, \text{ and}$$
$$TF = (\text{substantially}) \ 0.0$$

8. Conditions 13–18

a. SHAFT CONDITIONS (Fig. 19):

	Conditions 13–15	Conditions 16–18
Any shaft	Output	Input
Remaining shafts	Input and locked together	Output and locked together

b. SPEED AND TORQUE FACTORS. Reasoning such as that in Conditions 13–18 pertaining to the simple gear set shown in Fig. 11 shows that, in each of the above conditions,

$$SF = 1.0, \text{ and}$$
$$TF = 1.0$$

9. Summary and Graphs

The speed and torque factors in each of the conditions are indicated in Chart 5. Figure 20 includes graphs of the relationships between the factors in Conditions 1–6 and T_{s2}/T_{p2}, when

$$T_{p1}/T_{s1} = 1.0$$

Chart 5. SIMPLE PLANETARY GEAR SYSTEMS
(Simple Gear Set—Fig. 19)

Cond.	Shaft S_1	S_2	PC	SF
1	O	H	I	$1 - (T_{s2} T_{p1})/(T_{s1} T_{p2})$
2	I	H	O	$1/(1 - (T_{s2} T_{p1})/(T_{s1} T_{p2}))$
3	H	O	I	$1 - (T_{s1} T_{p2})/(T_{s2} T_{p1})$
4	H	I	O	$1/(1 - (T_{s1} T_{p2})/(T_{s2} T_{p1}))$
5	I	O	H	$(T_{s1} T_{p2})/(T_{s2} T_{p1})$
6	O	I	H	$(T_{s2} T_{p1})/(T_{s1} T_{p2})$
7–12	One	Free		$SF = 0.0 \quad TF = 0.0$
13–18	Two locked together			$SF = 1.0 \quad TF = 1.0$

Notes: (1) $TF = I/SF$ except in Conditions 7–12.
(2) SF and TF graphs in Fig. 20.

Other Simple and Compound Planetary Gear Systems

1. Conditions of Operation

The conditions of operation of other planetary gear sets, such as those shown in Fig. 21, in simple or compound gear systems correspond with those in the preceding chapters.

2. Speed and Torque Factors

The speed and torque factors of systems employing other gear sets may be computed by the general methods used in the preceding chapters. In some instances, the condition of operation and the factors correspond with those of another gear system. An example of this is the gear set shown in Fig. 21 D, which has the speed and torque factors indicated in the following.

a. CONDITIONS 1–6 (Fig. 21 D). These conditions and the corresponding simple system conditions of the gear set shown in Fig. 17 are:

TABLE 6. CONDITIONS 1–6—FIG. 21 D

	Shafts			Corresponding
Cond.	Sun 1	Sun 2	PC	*Condition in Chart 4*
1....	I	O	H	7
2....	O	I	H	8
3....	H	O	I	9
4....	H	I	O	10
5....	I	H	O	17
6....	O	H	I	18

b. Conditions 7–12 (Fig. 21 D). These conditions are those in which one shaft is Free. The SF and the TF are zero.

c. Conditions 13–18 (Fig. 21 D). These conditions are: any two shafts locked together and used as Input or Output; remaining shaft, Output or Input. The SF and the TF are 1.0.

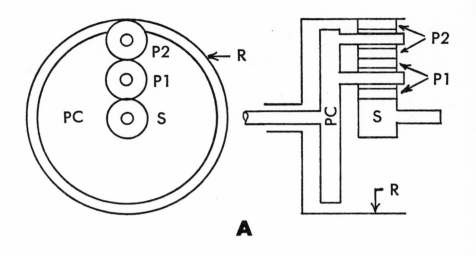

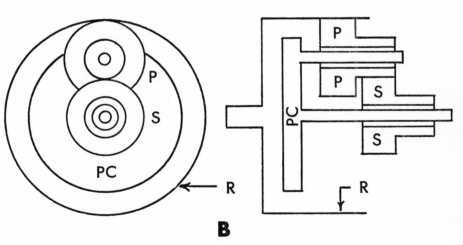

Fig. 21. Simple planetary gear sets. Only one planet pinion of each set shown.

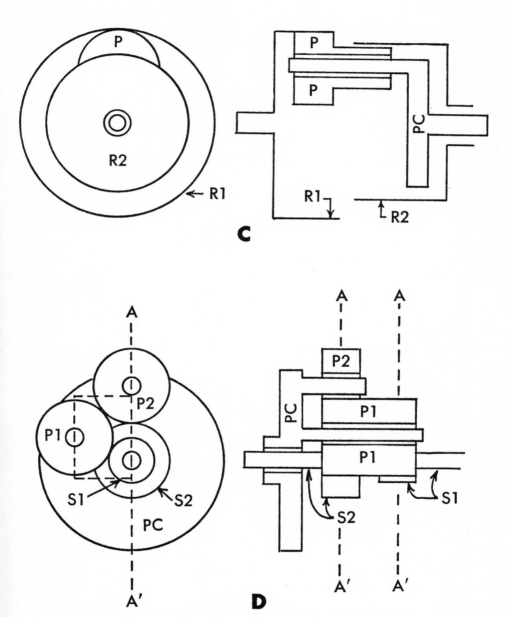

Fig. 21 (*Continued*). Only one planet pinion of each set shown.

103

10

Transmission Control Systems

1. General Description

The devices used in automatic transmission control systems are of two types. The first are those which are operated in a direct fashion by the driver. These are the lever or push-buttons used in placing the transmission in the Drive, Reverse, Park, etc., conditions, and the engine throttle control (accelerator) which not only controls the throttle but also affects the up-shift and down-shift actions of the transmission. They may include a control associated with the throttle control that changes the angle of the stator vanes in the fluid torque converter and thus changes the torque factor of the converter. The second type includes those controls which operate automatically when conditions are such that they should operate. These may include a control operated by the engine intake-manifold pressure (or vacuum) which is used in addition to the throttle control of the transmission. They usually include a control which operates in accordance with the automobile speed and causes the transmission to shift gears and thus change the torque and speed factors of the gear portion of the transmission. The transmission shown in Fig. 30 does not have this control.

Some of the above controls, in turn, control the fluid pressure that is applied to open and close gear-shift valves. These direct transmission fluid to piston-and-cylinder devices which engage and disengage clutches and bands in the transmission. These clutches or bands affect the TF of the gear portion of the transmission. Other controls may control electric currents which serve functions associated with the TF values, or with the engine starter, etc. In some instances, one control is used in opposition to another so that a change in the transmission occurs when one control force becomes greater than another. The reasons which cause this opposing-forces type of control and some other features of the control system to be used are discussed in this chapter.

The automatic and semiautomatic shifting of gears occurs only when the transmission control device is in a forward-drive position. When in the Reverse or Neutral position, the transmission remains in Reverse or Neutral gear, assuming that the engine is running in the case of the Reverse position. Likewise, when the control is in certain of the forward-drive positions (such as Low, First, or Second) the transmission shifts through a limited range of gear positions or does not shift. The transmission is in Neutral gear when the control is in the Park position, but the output shaft is locked to the transmission case. This in turn holds the automobile driveshaft and the automobile stationary.

2. Functions

The functions of the control system of an automatic transmission correspond with those of a driver in controlling the operations of a manually controlled transmission and mechanical clutch, in controlling the speeds of the engine that are associated with these operations, and in using his knowledge of the existing engine and automobile speeds and load on the engine in determining whether a shift of gears is desirable, and, if so, whether the shift should be an up-shift or a down-shift. Therefore, the control system of an automatic transmission, including the controls of the engine which are associated with the controls of the transmission, should perform its functions in a manner which is similar to the way in which a skilled driver controls a manually controlled transmission, the clutch associated with the transmission, and the engine. The similarity of the control functions makes the actions of the skilled driver worthy of study. His actions, then, form a basis for the determination of the functions of the control system of an automatic transmission.

Assume that the driver in the following discussion has a thorough knowledge of the factors which affect the performance of his automobile, the wear of its components, the fuel and oil consumption, and so on. Assume, also, that he desires that the wearing of parts and the fuel and oil consumption be kept at low values, but that he also demands high or maximum performance (acceleration, etc.) from the automobile when this performance is needed. Assume, further, that he has the ability to achieve all of the above.

Certain of the characteristics of the driver's automobile are illustrated in Fig. 22. The power-on torque factors of the transmission, which is similar to that shown in Fig. 3, are 2.5, 1.5, and 1.0 in 1st, 2nd, and 3rd gears, respectively. The axle gears and the sizes of the wheels and tires are such that the automobile speed is 20 mph for each 1,000 rpm of the engine when the transmission is in 3rd gear. In 2nd or 1st gear, the speed is 13.33 or 8 mph for each 1,000 rpm of the engine.

The values of the engine speeds and the corresponding automobile speeds in the following are assumed to be the optimum values when fuel consumption, performance, etc., are considered.

105

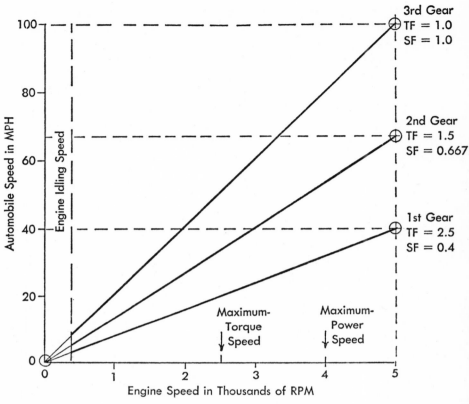

Fɪɢ. 22. Engine and automobile speeds.

a. MODERATE ACCELERATION ON A LEVEL ROAD. The driver disengages the clutch, places the transmission in 1st gear, adjusts the engine speed to a value between 750 and 1,000 rpm, and slowly engages the clutch. When the automobile has accelerated to 10 mph and the engine speed is 1,250 rpm, the driver disengages the clutch and shifts to 2nd gear, meanwhile completely or almost completely closing the throttle. The throttle closing prevents a high engine speed that otherwise would occur since there is no load on the engine when the clutch is disengaged. As the driver starts to engage the clutch he adjusts the engine speed to a value of 750 rpm. This speed permits the clutch to engage with no slipping if the automobile speed is still 10 mph, and prevents a sudden change in the speed of the automobile, either acceleration or deceleration, during the engagement of the clutch. After the clutch is engaged, the driver accelerates the automobile to a speed of about 17 mph, at which speed the engine speed is 1,250 rpm. He then repeats the above

106

process in shifting to 3rd gear, adjusting the engine speed to 850 rpm as he starts to engage the clutch.

b. FAST ACCELERATION ON A LEVEL ROAD OR MODERATE ACCELERATION ON A STEEP UP-GRADE ROAD. The driver controls the transmission, clutch, and engine during the above acceleration in the manner described for moderate acceleration on level roads except in one or possibly two respects. He accelerates the automobile to considerably higher speeds in 1st and 2nd gears before shifting to the next higher gear. These speeds may be as great as 30 and 55 mph, respectively, which speeds correspond approximately with the maximum-power speed of the engine. The second possible exception involves the speed of the engine during the clutch engagement. Instead of synchronizing the engine speed with the speed of the input shaft of the transmission, as in moderate acceleration on a level road, the driver may adjust the engine speed to a slightly higher value than that which would result in synchronization. Thus, torque for acceleration is provided by the engine during engagement of the clutch. This, of course, results in wear of the clutch, but the wear may be considered to be of secondary importance when compared with the need for maintaining or increasing the speed of the automobile during the clutch engagement. For similar reasons, an engine speed greater than 1,000 rpm may be used when the clutch is being engaged while in 1st gear.

c. FAST ACCELERATION AT NORMAL 3RD-GEAR SPEEDS OR OPERATIONS ON STEEP UP-GRADE ROADS. In this instance, the transmission is in the 3rd gear position, but acceleration of the automobile at a rate greater than the engine can provide in 3rd gear is required, or the axle torque needed to maintain the existing speed is greater than can be provided in 3rd gear.

The driver disengages the clutch, shifts to 2nd gear, and simultaneously increases the engine speed to a value greater than the speed before the shift in order to synchronize the engine speed with the transmission input shaft speed before the clutch is engaged. If he did not increase the engine speed to this value, there would be a sudden deceleration of the automobile, or "jerk," when the clutch is engaged. Alternately, the driver may increase the engine speed to a value greater than that required for synchronization and "slip" the clutch as he engages it in order to provide a forward torque on the driveshaft during the engagement. When the automobile speed has increased to, say, 60 mph, he shifts to 3rd gear since the torque of the engine is decreasing as the engine speed increases, the maximum-engine-power speed has been exceeded, and the strain on the engine because of its speed is greater than the driver desires it to be.

d. DECELERATION OR OTHER ENGINE-BRAKING OPERATIONS. In these instances, the driver closes the throttle and uses the power that is required to drive the engine at speeds greater than its idling speed to decrease the automobile speed or to assist in maintaining it at a reasonable value when descending hills. If more power is required for this than is provided by the automobile with the transmission in 3rd gear, the driver shifts to 2nd gear and possibly later to 1st gear.

The actions of the driver are the same as when shifting down to provide increased acceleration except possibly in one respect. In accelerating, the driver may adjust the engine speed to a speed greater than that required for synchronization with the transmission input shaft in order to provide a forward torque during the engagement of the clutch. In decelerating, he may adjust the speed to a lower value in order to provide a reverse torque on the driveshaft during the engagement.

e. ALL OF THE ABOVE OPERATIONS. The actions taken by the driver in all of the above and similar operations may be classified as:

(1) Using his knowledge of the speeds of the automobile and engine, the torque load on the engine, and other such factors, in determining which gear shift should be made and when it should be made.

(2) Shifting the gears of the transmission so as to provide a greater or lesser transmission torque factor and a corresponding but opposite change in the speed factor. This is accomplished by moving the transmission control lever from one of its positions to the position which provides the desired torque and speed factors.

(3) Removing the torque from the gears during the period of time the torque and speed factors are being changed, which prevents undue stresses on and possible damage to the gears. This is accomplished by disengaging the clutch.

(4) Preventing high and unneeded engine speed during the time the engine is not coupled to the automobile driveshaft by the transmission and the clutch.

(5) Applying the torque of the engine to the driveshaft, or the torque of the driveshaft to the engine, after the change in torque and speed factors is made. This is done by engaging the clutch.

(6) Synchronizing the engine speed with the speed of the input shaft of the transmission, or with the speed of the "new" output shaft in power-off operation. This may be done prior to engaging the clutch or during the engagement.

The last action is accomplished by the use of the throttle or clutch, or both. The speed of the input shaft of the transmission is zero when the automobile is stationary, the gear-shift lever is in the 1st gear or Reverse gear position, and the clutch is disengaged. The speed of the engine, if running, is at least its idling speed, or 400 rpm in the case of the automobile shown in Fig. 22. Usually, an engine speed greater than the idling speed—say, 750 rpm or more—is desirable in starting the automobile in order that the engine will not "die" during the starting. The only available means of synchronizing the engine and the transmission input shaft speeds is the clutch. The driver engages it slowly so that the torque applied to the input shaft does not increase so rapidly that it causes the automobile to "jerk" forward (or in the reverse direction if the transmission control is in the Reverse position). This slow engagement also prevents the torque requirement which is placed on the engine from being so great as to "kill" the engine.

In shifting from 1st to 2nd and from 2nd to 3rd gears, or vice versa, it is possible to obtain synchronization by the use of the throttle only. However, even a skilled driver probably accomplishes this in only a small percentage of the shifts he makes. In fast acceleration or deceleration operations, as indicated above, he purposely may use the clutch and throttle in obtaining the synchronization.

The shifting-up of gears is done when the engine speed after the shift will be such that the engine can produce the torque required under the operating conditions. This may occur at a relatively low or at a relatively high automobile speed. In the example of fast acceleration at normal 3rd gear speeds above, the up-shift from 2nd to 3rd gear occurred at 60 mph. Down-shifting is done when the engine torque is inadequate to provide the desired performance in the higher gear in power-on operations, or when the driveshaft cannot transfer the desired amount of power to the engine for braking purposes in power-off operations.

The prevention of high and unneeded engine speed during shifts may be accomplished in one or both of two ways. The engine may be throttled during the time it is not coupled to the driveshaft, or the gear-shifting cycle may be accomplished so rapidly that the engine does not have enough time to accelerate to an unneeded high speed. A modified fast-shifting method is used to a considerable extent when shifting down, in which cases the engine is not throttled and its speed is allowed to increase to the value it will have after the shift is completed.

3. Automatic Transmission Controls

The proper timing of gear shifting in power-on operations involves two principal factors: (1) the engine speed and (2) the torque load that is placed on the engine as the result of the speed and/or desired acceleration of the automobile and of the up-grade, down-grade, or level and other conditions of the road. These factors may vary widely and in opposite fashions, or widely and together. For example, the engine speed may be high and the torque load light when driving on down-grade roads, or the engine speed may be low and the torque load great on up-grade roads.

The correlation of speed and torque load factors might be obtained (with the limitations indicated below) by using two opposing forces in

109

the control system, one of which is controlled by the speed of the engine and the other by the torque load on the engine. In practice, the speed of the automobile rather than the speed of the engine is used to control the first gear-shift control force. The automobile speed is a function of the engine speed, the speed factor of the gear portion of the transmission, and the speed factor of the fluid clutch or torque converter in the transmission. If the torque required by the automobile is great, the latter speed factor is lower than when the required torque is small. Therefore, for any given engine speed, the transmission control force which results from the speed of the automobile is less under heavy-torque requirement conditions than under light-torque conditions. Thus, the control force is a function of the load on the engine as well as of the engine speed. If engine speed rather than automobile speed were used, the transmission might shift from 1st gear to 2nd gear when the engine speed reaches a particular speed even though the output shaft of the transmission is stalled and the automobile is stationary. The gear-shift control force which is a function of automobile speed is used to cause up-shifts from 1st to 2nd and from 2nd to 3rd gears in three-speed transmissions and the corresponding shifts in two-speed or four-speed transmissions. Likewise, it is used to prevent down-shifting.

The gear-shift control force which is based on the torque load of the engine is controlled partially or wholly by the position of the throttle control (accelerator). The degree of opening of the throttle is interpreted by the control system as being an indication of the torque load on the engine. In some control systems, the engine intake-manifold pressure (or vacuum) is used in addition to the throttle control in determining the value of the gear-shift control force since the value of this pressure is an indication of the torque load on the engine. The gear-shift control force based on engine load is used to prevent up-shifts and to cause down-shifts.

The above two gear-shift control forces are provided by one or both of two transmission fluid pumps. The gear-shift control force which, in turn, is controlled by the automobile speed is provided by a pump that is driven by the output shaft of the transmission. A pressure-regulating valve associated with an output of this pump is operated by a governor type of control in accordance with the speed of the transmission output shaft. At low automobile speeds the fluid pressure is small and at high speeds it is large. The gear-shift control force which is controlled by the engine torque is provided by a fluid pump that is driven by the engine, or under some operating conditions by the pump that is driven by the transmission output shaft. The fluid pressure which provides the force is governed by the throttle control or intake-manifold pressure, or both. The pressure increases as the throttle opening increases and as the mani-

110

fold pressure increases. This pressure usually is supplemented by spring pressure.

The two gear-shift control forces act in opposition to one another, first in controlling the 1st–2nd gear-shift valve, assuming the transmission is a three-speed transmission operating in the Drive position. After (and perhaps before) the shift to 2nd gear is made, the forces act in opposition on the 2nd–3rd gear-shift valve. In each case, when the force acting to cause an up-shift is greater than that acting to prevent a shift, the shift valve concerned operates and fluid pressure is placed on piston and cylinder devices which cause engagement of the clutches and bands in the gear part of the transmission that are engaged in the next higher gear. When the force acting to cause a down-shift is greater than that acting to maintain the existing gear ratio, the gear-shift valve (or valves) concerned operates and fluid pressure engages the clutches and bands which are used in a lower gear, which may or may not be the next lower gear. The values of the forces which tend to cause down-shifts and up-shifts are made such that certain automobile speeds cannot be exceeded in 1st and 2nd gears when in the Drive position. This is done for the same reasons that caused the skilled driver of the manually controlled transmission automobile to shift to 3rd gear at 60 mph in the example of fast acceleration at normal 3rd-gear speeds.

As stated above, the transmission shifts gears when the force tending to cause a shift is greater than that tending to prevent a shift. Let it be assumed that a two-speed transmission is designed and constructed in such a manner that it shifts from 1st to High gear when the speed of the automobile is 15.1 mph if the throttle is slightly open, and down-shifts to 1st gear when the speed is 15 mph and the throttle setting is the same. Assume, further, that these shifts occur at 16.1 and 16 mph if the throttle opening is 5 per cent greater than in the above, and at 14.1 and 14 mph if the opening is 5 per cent smaller.

When driving in some traffic or other conditions, it may be desirable to maintain a speed in the range of 13 to 16 mph. The road conditions may be such that slightly different throttle settings are needed from time to time to maintain any one speed in this range. Consider the situation in which a speed of 15.2 mph is desired and in which the road conditions vary slightly. The speed of the automobile will vary from, say, 14.5 to 15.5 mph if the driver maintains the first throttle setting above and if the transmission remains in High gear. However, if the speed decreases to 15 mph, the transmission shifts to 1st gear. The decreased transmission SF resulting from the shift probably requires a greater throttle opening until the speed of 15.2 mph is reestablished. If the driver moves the throttle to provide a 5 per cent greater opening in order to prevent the above decrease in speed from 15.2 to 15 mph, a down-shift occurs be-

111

cause the speed is less than 16 mph. Similar unneeded shifting may occur if the desired speed is in the vicinity of 14.5 mph. Thus, the control system "hunts" in order to maintain the relationships between speed, throttle setting, and gear position that were used as bases in the design of the transmission. This repeated and unnecessary shifting is objectionable from the standpoints of the driver and of wear of the transmission parts. Therefore, the designs are so made that unnecessary shifting is eliminated or greatly reduced. This may be accomplished by making the first shift points above, say, 15 and 9 mph instead of 15.1 and 15 mph. Likewise, the shift points with the greater and lesser throttle openings might be 16 and 10, and 14 and 8 mph. Then a considerable percentage change in speed is needed to produce a down-shift when the throttle setting remains constant, and a considerable change in throttle setting is needed if the speed remains constant.

The large difference between the up-shift and down-shift points for any given throttle setting can be obtained by the use of a gear-shift valve which controls a passageway(s) to a piston which provides part of the throttle control force acting on that valve. When the valve moves to cause an up-shift, it opens a passageway after moving a small amount. Fluid flows through the passageway and removes the pressure from the piston. This decreases the total force which tends to hold the valve in the lower gear position since the total area on which the fluid exerts pressure is less than when completely in the lower gear position. As a result, the gear-shift valve moves to the higher gear position. Then, when the automobile speed remains constant, a greater fluid pressure per square inch from the throttle control is needed to cause a down-shift than would be needed if the pressure area were not reduced. Likewise, less fluid pressure per square inch is needed from the automobile-speed control to keep the transmission in High gear if the throttle setting remains constant. This method of lowering the down-shift speeds has an additional advantage. When the gear-shift valve opens the passageway and removes the pressure acting on the piston, the reduction in the throttle force acting on the valve causes the valve to move quickly and firmly into the higher gear position. Without this reduction in force, the valve may move slowly, or move back and forth in accordance with variations in the control forces acting on it. The clutches and bands in the gear system, then, may engage and disengage slowly and in an unpredictable manner, or they may tend to vary in the extents of engagement from not engaged to partially engaged to not engaged, and so on.

In some transmissions, a change in the torque factor of the torque converter is made in lieu of or in addition to a change in the torque factor of the gears. The change to a higher converter torque factor is made when the throttle is opened more than a particular amount—for

112

example, more than three-fourths open. Conversely, the same change may be made in the converter when the throttle is closed. This reduces the transmitted torque, thereby reducing the tendency of the automobile to "creep" when the engine is idling. (See Chapter 2, § 6.)

The above actions of the transmission control system correspond with actions (1) and (2) of the driver of the manually controlled transmission automobile, as listed in § 2, e, above. The other actions of the driver are very closely related. In an automatic transmission, the fluid unit which replaces the mechanical clutch in the manually controlled system always is in an engaged condition. However, the extent to which it is engaged—that is, the extent to which it can transmit torque—is dependent on the speeds of the input and output shafts of the unit. Therefore, it could be said to be lightly engaged when the engine is idling and the automobile is stationary. The partial engagement and the resultant slipping of the fluid unit does not cause wear of the unit in the sense that slipping a mechanical clutch causes wear. Also, an increase in the degree of engagement of a fluid unit causes much less "jerk" of the automobile than is apt to be caused by a mechanical clutch. For these reasons, a fluid unit can be used in synchronizing the engine speed and the speed of the input shaft of the gears of the transmission to a greater extent than a mechanical clutch can (or should) be used. The function of applying torque to the driveshaft in power-on operations, or to the engine in power-off operations, after a gear-shift change is made is performed automatically and in general smoothly by a fluid unit, and by manual control of the throttle in some instances.

The gears used in automatic transmissions usually are of a planetary type, in which the gears always are in mesh. However, stresses are placed on the clutches, bands, and other parts of the gear portion of the transmission if the torques on them and the gears are great during shifts, and wear of these parts results from the stresses. The action of the fluid unit limits these stresses and wear, but additional measures are desirable to reduce them and to reduce strong momentary acceleration or deceleration of the automobile which might be called "surges," instead of "jerks" as they were called in the discussion of the manually controlled transmission. This change in nomenclature is intended to indicate that the acceleration or deceleration in the automatic transmission case is strong but smooth, while those in the manually controlled transmission case may be of shorter duration and abrupt.

The synchronization of the engine speed with that of the input shaft of the gears by an automatic throttle control probably is impractical in automatic transmissions unless a longer period of time is employed than is desirable between disengagement of the clutches, etc., associated with one gear ratio and the engagement of those associated with another. A means of controlling the engine speed has been used in some nonplanetary gear transmissions but more for the purpose of removing torque from the gears during shifts than for synchronization. This involves the interruption of the engine ignition system during the short time period of the shift. A more commonly used method of reducing stresses and surges in power-on up-shifts involves devices in the fluid passages which cause the clutches, etc., pertaining to one gear ratio to disengage simultaneously with the engagement of the corresponding clutches, etc., pertaining to the next-higher gear ratio. This prevents an appreciable increase in the engine speed during an up-shift. Also, completion of the engagement of the clutches, etc., may be made slowly so that a quick change in the torque does not occur. This is analogous to slipping the clutch during its engagement when shifting the gears of a manually controlled transmission. In power-on down-shifts, slow (and/or delayed) engagement permits the speed of the engine to increase to the value needed upon completion of the shift. In a similar manner, this engagement permits the engine speed to decrease in closed-throttle up-shifts. These slow and/or delayed engagements are provided by such means as restrictions in the fluid passageways which may be by-passed by a valve if their use is not desired in other shifts. An accumulator may be used in conjunction with a fluid restriction. It includes a piston in a cylinder and a spring which tends to keep the piston at one end of the cylinder. In some instances, the accumulator is connected to the fluid passageway between the restriction and the piston-and-cylinder device that engages a band or clutch. When the gear-shift valve opens the passageway from the fluid pump, the fluid flows through the restriction to the accumulator and to the band or clutch engagement device. The pressure of the fluid moves the accumulator piston against the spring force and in doing so increases the volume of fluid in the accumulator. The pressure in pounds per square inch on the piston of the band or clutch engagement device at any instant, then, is the same as that on the accumulator piston. The pressure increases at a rate which is determined by the size of the opening in the restriction, the force exerted by the spring vs. movement of the accumulator piston, and the volumes of the accumulator cylinder and the cylinder of the band or clutch engagement device. The extent of the engagement of the band or clutch varies with this pressure, so the rate of the engagement varies in accordance with the rate at which the pressure changes. An accumulator also may be used in controlling the rate of disengagement of a band or clutch. When the gear-shift valve closes the passageway to the accumulator and engagement device and the fluid is allowed to escape from their cylinders, the pressure per square inch within them decreases at a rate which is determined by the size of the fluid-escape passageway, the force of the accumulator spring, and the sizes of the cylinders. An accumulator may be more complex than the one described above, and fluid pressures may be applied to both sides of the piston.

The transmission control system of an automatic transmission, then, performs its functions in a manner and to an extent very similar to the performance of functions by a skilled driver of an automobile which has a manually controlled transmission. It can be argued that a skilled driver has a greater capability of performing these functions correctly than the control system of an automatic transmission. This may be true from the standpoint of synchronism of the engine speed with the speed of the input shaft of the gears if it is assumed that the skilled driver's best performance in this respect is equalled or nearly equalled in each succeeding shift of gears. This assumption may not be an acceptable one. If the performance of all control functions during several thousand gear shifts by an automatic transmission and the performance during a like number of shifts by a skilled driver using a manually controlled transmission were compared, it is probable that the performance of the automatic control system would be found to be the equal of that of human control. If the same comparison is made but with a representative group of drivers experienced in the use of manually controlled transmissions but with skills ranging from poor to superior, it is probable that the performance of the automatic control system will be found to exceed considerably the average of the performances of the drivers of the manually controlled transmission automobiles.

4. Effects on Torque and Speed Factors

a. FACTORS OF FLUID UNITS. As stated before, the control system changes the torque factor of the torque converter in some transmissions. In the transmission shown in Fig. 33, the control system controls the torque capability of one of the fluid clutches and thus changes the torque and speed factors of the clutch and its associated gear set.

In Chapter 2, § 5, it is stated that the torque factor of a fluid clutch is less than 1.0, but that normally it is slightly less than 1.0. In § 6 of that same chapter, similar statements are made concerning torque converters in those instances in which the stator rotates freely. This occurs when the slip is small and the impeller is the input element, and when the slip has any normal value and the turbine is the input element. The small difference between 1.0 and the *TF* of a fluid clutch, and between 1.0 and the *TF* of a converter under the above conditions is disregarded in Chapters 12 to 20 pertaining to automatic transmissions of various types, so the *TF* values used in those chapters are 1.0.

Likewise, it is stated in Chapter 2 that the *SF* of a fluid clutch or torque converter which is transmitting torque always is less than 1.0, but that it is nearly 1.0 when the load presents a torque to the output shaft of the clutch or converter which can be transmitted with a small

115

slip. In Chapters 12 to 20, the ranges of speed factors are based on the assumption that the speed factor is substantially 1.0 when the slip required to produce the output torque is small.

b. POWER-ON OPERATION. Power-on operations are those in which the engine drives or tends to drive the automobile driveshaft. In the case of operations in a gear which is not the lowest gear available with the transmission control in a particular position (such as 2nd or 3rd gear when the control of some transmissions is in the Drive position), the torque and speed factor ranges may be stated in Chapters 12 to 20 in such forms as

$$TF_{\text{trans}} = 1.5 \text{ to a higher value, and}$$
$$SF_{\text{trans}} = 0.667 \text{ to a lower value}$$

The higher and lower values may be those that exist when the control system shifts the transmission to a lower gear (the next lower or a lower gear) because the throttle control force exceeds the automobile-speed control force, assuming the automobile speed to be below a certain speed. If it is above that speed, the SF of the transmission has a value that is determined by the speeds of the automobile driveshaft and the transmission input shaft, which in turn are determined by the automobile speed and the throttle setting. The slip of the fluid unit is $(S_{\text{in}} - S_{\text{out}})/S_{\text{in}}$, and this determines the TF of the unit in the case of torque converters, and thereby determines the TF of the transmission.

The values of the TF and SF also may be those which exist when the control system shifts the transmission from 2nd or 3rd gear to a higher gear as the result of the automobile-speed control force being greater than the engine-torque control force.

c. POWER-OFF OPERATION. Power-off operations are those in which the automobile drives or tends to drive the engine at a speed which is greater than it would be at the existing throttle setting if the engine were not coupled to the driveshaft. Two of the possible types of power-off operations are considered: (1) operation on a level or up-grade road; (2) operation on a steep down-grade road. In either of these, the engine ignition may be on or off.

In Chapters 12 to 20, the torque and speed factors pertaining to Reverse gear and to a forward gear when the transmission control is in a position in which the transmission does not shift gears may be stated in forms such as

$$TF_{\text{trans}} = 0.454, \text{ and}$$
$$SF = 2.2 \text{ to } 0.0$$

In these instances, the 0.0 and near 0.0 speed factors pertain to level or up-grade roads and to ignition-off operations, or to "push-to-start" or

116

similar operations, since otherwise the engine would drive the "new" output shaft of the transmission at low or zero automobile speeds and the "new" output shaft speed would not be the result of the speed of the "new" input shaft. In operations on steep down-grade roads, the automobile continues to drive the engine, assuming that the brake is not used to stop the automobile or to decrease its speed to a low value. In these cases, the SF of the transmission is determined by the existing engine speed and automobile speed. These, together with the SF of the transmission gearing, determine the value of the slip of the fluid unit, which, in turn, is that which is required to provide the torque to drive the engine at its speed. (This torque has been discussed in preceding chapters.)

When the transmission control is in a position in which the control system shifts the gears, the factors may be expressed in forms such as

$$TF_{\text{trans}} = 0.454, \text{ and}$$
$$SF_{\text{trans}} = 2.2 \text{ to a lower value}$$

The SF value is that which exists when the automobile has a speed at which the control system shifts the transmission to a higher or a lower gear.

When shifting to a lower gear in power-off operation, the shift may not be to the next-lower gear. For example, the control system of a three-speed transmission may cause a shift from 3rd to 1st gear when the automobile speed is approximately 10 mph and the throttle is closed.

d. Shift Patterns. The pattern of the gear shifts that are made by a particular type of transmission may be shown in summary form as in Table 7. This table is representative of those of three-speed transmissions which shift throughout the range of the three speeds when in the Drive position. Ranges of automobile speeds usually are specified in shift-pattern tables, rather than particular speeds, since the speed at which a shift occurs varies in transmissions of the same type due to such factors as variations in the operating characteristics of the fluid valves, the condition of the engine, etc. The pattern in Table 7 includes ranges of speeds and "nominal" specific speeds which are intended to indicate the relative values of the speeds in a particular well-adjusted transmission. In Kick-Down operations, the accelerator is pushed all of the way down for the purpose of causing a down-shift to occur. These do not occur at speeds in excess of those listed.

When the throttle control is between the closed-throttle and the full-throttle positions, the shift speeds change accordingly. For example, when the throttle is in the one-quarter-open position, the up-shifts from 1st to 2nd gear and from 2nd to 3rd gear may occur at approximately 20 and 30 mph.

TABLE 7. TRANSMISSION SHIFT PATTERN

Conditions and Shifts	Ranges of Speeds (mph)	Nominal Speeds (mph)
Closed throttle:		
1st–2nd up-shift	7–15	11
2nd–3rd up-shift	13–20	17
2nd–1st down-shift	6–14	8
3rd–1st down-shift	6–14	8
Full throttle:		
1st–2nd up-shift	35–50	42
2nd–3rd up-shift	75–85	80
Kick-down (upper limits):		
3rd–2nd down-shift	60–80	70
2nd–1st down-shift	30–40	35
3rd–1st down-shift	30–40	35

11

Model T Ford Transmissions

1. General Description

The Model T Ford transmission is a manually controlled transmission which employs a planetary gear set without a ring gear. Figure 23 illustrates a transmission which is the same as or similar to that of the Model T Ford. The gear set is of the general type shown in Fig. 19, but it is a compound set rather than a simple one.

The transmission is included in this discussion of automatic transmissions for two reasons:

(1) The above type of gear set, while not used in existing American

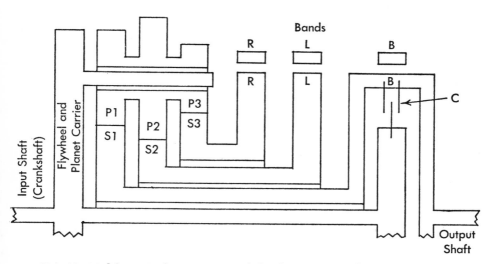

Fig. 23. Model T Ford transmission. (1) Three compound pinions are used. (2) R, L, and B indicate the Reverse and Low-gear bands and drums; C, the High-gear clutch. (3) Bands are connected to transmission case.

automatic transmissions, may be used in the future, and the plan of use may resemble that shown in Fig. 23.

(2) The transmission could be employed as an automatic transmission if automatic control devices were added to it.

The use of a fluid clutch or torque converter between the engine and the gears would be desirable if the transmission were converted into an automatic transmission.

2. Transmission Controls

The transmission has four controls, three of them being foot pedals and the fourth a hand lever. The hand lever serves as the parking brake control, but when it is placed in the brake-on position it moves the Low-High-Neutral pedal (the left foot pedal) to its half-down or Neutral-gear position. It does not lock the transmission output shaft to the transmission case or other object. The normal position of the center pedal, the Reverse-gear pedal, is behind those of the others. This position tends to prevent errors by the driver in placing his feet on the pedals. The control positions and the corresponding gear operations are shown in Table 8.

TABLE 8. TRANSMISSION CONTROL POSITIONS

Gear	Low, High, and Neutral Pedal	Reverse Pedal	Brake Pedal	Parking Brake Lever
Low	Down	Up	Up	Forward
High	Up	Up	Up	Forward
Reverse	Half-down	Down	Up	Forward
Neutral	Half-down	Up	Up or down	Forward, or back if parking

The brake pedal may be pushed partially down in Low-, High-, or Reverse-gear power-off operations when deceleration of the automobile is desired, or to prevent acceleration on down-grade roads. If pushed completely down, it stops the rotation of the output shaft and therefore the input shaft which is the engine crankshaft, thus stopping the engine. For this and other reasons, the use of the brake pedal normally is restricted to High-gear power-off operations and to Neutral-gear operations.

3. Numbers of Gear Teeth

The numbers of teeth on the gears are assumed to be:

120

$$T_{p1} - 27 \qquad T_{s1} - 27$$
$$T_{p2} - 33 \qquad T_{s2} - 21$$
$$T_{p3} - 24 \qquad T_{s3} - 30$$

4. Condition of Operation in Low and Reverse Gears

Condition 1 of the simple gear systems in Chart 5 exists in Low-gear and Reverse-gear operations when the engine provides driving torque to the transmission. The speed factor in this condition is

$$SF = 1 - (T_{s2}\, T_{p1})/(T_{s1}\, T_{p2}), \text{ and}$$
$$TF = 1/SF$$

5. Low-Gear Power-On Operation

In Low gear the L band Holds sun gear 2 and all other bands and the clutch are disengaged.

Substituting the gear teeth values in the SF and TF equations pertaining to Condition 1,

$$TF_{\text{trans}} = Tq_{\text{out}}/Tq_{\text{in}} = 2.75, \text{ and}$$
$$SF_{\text{trans}} = S_{\text{out}}/S_{\text{in}} = 0.364$$

6. Reverse-Gear Power-On Operation

In Reverse gear the R band Holds sun gear 3 and all other bands and the clutch are disengaged. The conditions are the same as in Low gear if T_{p3} and T_{s3} are substituted in the SF and TF equations in place of T_{p2} and T_{s2}. The speed and torque factors are

$$TF_{\text{trans}} = -4.0, \text{ and}$$
$$SF_{\text{trans}} = -0.25$$

The output shaft rotates in the direction opposite to that of the input shaft (the crankshaft), or in the reverse direction.

7. High-Gear Power-On Operation

In High gear the clutch is engaged and all bands are disengaged. The planet carrier and sun gear 1 are locked together. Therefore, the planet pinions do not rotate on their shafts. One of Conditions 13–18 in Chart 5 exists and the TF and SF of the transmission are 1.0.

8. Neutral-Gear Operation

In the Neutral-gear position, the L and R bands and the clutch are disengaged. The B band may be engaged or disengaged. Brake shoes not shown in the illustration may be forced against drums attached to the rear wheels of the automobile. These shoes are part of the parking brake system.

121

Torque from the engine rotates the flywheel and planet carrier. The planet pinions rotate around the axis of the flywheel shaft. They also rotate around their own axes since sun gear 1 is connected (indirectly) to the rear wheels of the automobile and therefore tends to be stationary. There is no force, other than a relatively small force due to friction, opposing the rotation of the L and R drums. Therefore, these drums rotate freely and substantially no torque is applied to sun gear 1 and the output shaft of the transmission. The automobile remains stationary if the torque required to drive the rear wheels is appreciable, including the torque required to overcome the effects of the rear-wheel brakes if these are applied. One of Conditions 7–12 in Chart 5 exists and the SF and TF of the transmission are 0.0, or substantially so.

9. Braking Operations

In Braking operations, the B band partially or completely Holds its drum, and the clutch may be engaged or disengaged. Normally, it is disengaged when the automobile has decelerated to a low speed. Other bands normally are disengaged prior to applying the brake if they are applied, as in Low- or Reverse-gear operation.

The tightening of the B band on its drum decreases the speed of the output shaft, which in turn is connected (indirectly) to the rear wheels.

10. Power-Off Operation

With one exception, the clutch and band conditions in the various gears when the automobile is operating with the engine throttled to a power-off state are the same as the conditions when operating with power on. The exception is the B band, which may or may not be engaged with its drum. If it is engaged, it serves to reduce the torque that is applied to the gear portion of the transmission but does not affect the operation of that portion in any other way.

The output shaft shown in Fig. 23 becomes the "new" input shaft of the transmission and the input shaft becomes the "new" output shaft in power-off operations. When the functions of the input and output shafts are exchanged, the conditions of operation of the gear set change accordingly, and the TF and SF values are the reciprocals of those that existed before the exchange of functions, except in Neutral gear. Therefore, these values become the following when the TF effects of the transmission brake are negligible—that is, when this brake is not applied. The rear wheels drive the engine in all gears except Neutral gear and the braking effect of the engine friction assists in decelerating the automobile. Chapter 1, § 3, and Chapter 2, § 1 include discussions of the braking effect of engine friction.

TABLE 9. POWER-OFF SPEED AND
TORQUE FACTORS

Gear	SF	TF
Low	2.75	0.364
High	1.00	1.000
Reverse ..	—4.00	—0.250
Neutral ..	0.00	0.000

11. Control System Considerations

The transmission shown in Fig. 23 and the transmission and clutch shown in Fig. 3 differ greatly in their construction, but the functions of drivers in controlling them and the engines correspond closely with one another. In starting an automobile in the forward or reverse direction, the L or R band in Fig. 23 is engaged slowly in order to prevent the sudden placing of a large torque requirement on the engine and the sudden application of large torques on the parts of the transmission and on the driveshaft. Nevertheless, a drum may transmit relatively large torques to its band throughout the engagement of the band, as shown in the following.

When the automobile is being started in Reverse gear, the TF of the transmission is —4.0. Using relationships corresponding with those in Chapter 5, § 2,

$$Tq_{\text{in pc}} + Tq_{\text{in s1}} + Tq_{\text{in s3}} = 0.0, \text{ or}$$
$$Tq_{\text{in trans}} - Tq_{\text{out trans}} = Tq_{\text{out R drum}}, \text{ or}$$
$$Tq_{\text{out R drum}} = Tq_{\text{in trans}} + 4.0\, Tq_{\text{in trans}} = 5.0\, Tq_{\text{in trans}}$$

This value of the output torque of the R drum also may be computed by assuming that the load on sun gear 1 shaft Holds that shaft and considering the sun gear 3 shaft as being the output shaft. The speed of the R drum immediately prior to the engagement of the band is

$$S_{\text{R drum}} = (S_{\text{in trans}})\,(SF \text{ in Condition 3, Chart 5})$$

or 200 rpm if the engine speed is 1,000 rpm. This speed is reduced to zero when the band is completely engaged.

When the automobile is being started in Low gear,

$$Tq_{\text{out L drum}} = Tq_{\text{in trans}} - 2.75\, Tq_{\text{in trans}} = -1.75\, Tq_{\text{in trans}}$$

The speed of the L drum immediately prior to the engagement of the band is —571 rpm if the engine speed is 1,000 rpm.

The Reverse or Low drum torque and speed, both of which vary in

123

magnitude during the engagement of the band as the speeds of the automobile and engine vary, represent power which is dissipated in the form of heat by the band and drum. The temperatures of the band lining and drum and the abrasive action which results from the slipping of the band on its drum are unfavorable factors in the useful life of these items, particularly the lining.

Skillful engagement of either of the bands involves judgments of the torque requirement of the automobile and of the torque capability of the engine at each instant, and adjustment of the throttle control (a hand lever rather than a foot pedal) or band engagement (or both) so as to maintain a proper relationship between torque requirement and capability.

A skillful shift from Low to High gear involves a coordination of the throttle control and clutch engagement, and a shift from High to Low gear involves a coordination of the throttle and Low band engagement. These coordinations are similar to those discussed in Chapter 10 in connection with the shifts of the transmission in Fig. 3 from one gear to another.

Let it be assumed that the transmission shown in Fig. 23 is to be controlled by an automatic control system. The controls which cause the engagement of the Low and Reverse bands would need to be governed by the relationship between the torque requirement and capability above. Thus, the system might be designed so that:

(1) No engagement occurs when the throttle is in the idling-speed or near idling-speed position,

(2) The degree of engagement in another throttle position increases with time until the engagement is complete, unless the difference between the existing speed of the engine and the speed it would have with no load becomes greater than certain values.

The no-load speed of the engine is a function of the throttle setting, so this setting could be used to control one of the band-engagement control forces. The controls of the shifts from Low to High gear, and vice versa, might be the same as those described in Chapter 10.

The control system above has certain objectionable features. First, the Low or Reverse band continues to slip as long as the difference between the existing speed and the no-load speed of the engine exceeds certain values. This may result in excessive temperatures and wear of the bands and drums. Second, the shifts from Low to High gear and vice versa may result in sudden large torques on the engine, transmission parts and driveshaft unless the completions of the engagements of the Low band and High gear clutch are made slowly since the throttle setting normally is not changed during automatic shifts and since there is no torque-cushioning device in the system other than the clutch or the band and drum.

The use of a fluid clutch or torque converter in the torque train would eliminate the need for the above control of the band engagement in starting the automobile, and would reduce the torques which result from the changes in speed factor during shifts from Low to High gear, and vice versa. These torques are produced by the kinetic energy of the engine or automobile. The elimination of the control which prevents engagement of a band when the throttle is closed or nearly closed would cause the TF of the transmission in Low-gear and Reverse-gear power-off operations to be 0.364 and —0.25, rather than 0.0 in each case. For reasons such as these, the use of a fluid unit would be desirable if automatic control were used. It also would provide benefits when manual control is used.

12

TorqueFlite Transmissions

1. General Description

TorqueFlite transmissions employ a lever, or push-buttons and a lever, as manual control devices. The throttle control also serves as a manual transmission control, as described in Chapter 10. The manual and automatic controls cause the transmission to shift gears in the ways discussed in that Chapter, and as indicated below. A dual planetary gear system is used in the transmissions.

The transmission control (or controls) places the transmission in any one of six conditions: Drive, Second, First, Neutral, Reverse, and Park.

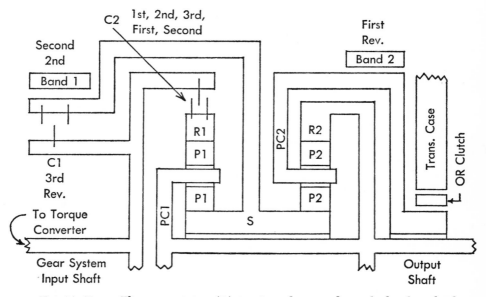

FIG. 24. TorqueFlite transmission. (1) 1st, Second, etc., indicate the band or clutch is engaged in 1st gear, etc. (2) *OR* clutch Holds when planet carrier 2 tends to rotate in reverse direction. (3) Bands are connected to transmission case.

126

When the control is in the Drive position, the transmission shifts from 1st to 2nd to 3rd gear, and vice versa or from 3rd to 1st gear. The 3rd-to 1st-gear shift occurs at about 10 mph in closed-throttle operations, or at speeds below approximately 35 mph when the accelerator is pushed completely down—that is, pushed to the Kick-Down position. When in the Second position, the transmission shifts from 1st to Second gear, and vice versa. When in the First, Reverse, Neutral, or Park position, the transmission remains in First, Reverse, or Neutral gear. The terms 1st gear, First gear, etc., and the conditions when in the Park position are discussed in Chapter 2, § 4.

Figure 24 illustrates a transmission which is the same as or similar to a TorqueFlite transmission. It also is the same as or similar to a C4 Automatic Dual Range transmission. The transmission includes an overrunning clutch which is described in Chapter 2, § 6. The three-element torque converter in the power train between the engine and the gear system has a maximum torque factor of approximately 2.0.

Chapter 10, § 4, provides information pertaining to the TF and SF values as they are stated in this chapter.

2. Numbers of Gear Teeth

The numbers of teeth on the gears of the transmission are assumed to be:

Sun gear — 28
Planet pinions 1 and 2 — 17
Ring gears 1 and 2 — 62

In several instances in the following, the sun gear is treated as though it were two sun gears, one for each of gear sets 1 and 2.

3. 1st-Gear Power-On Operation (In Drive or Second Position)

Clutch 2 is engaged.

The power flow is from the input shaft of the gear system to the output shaft through gear sets 1 and 2. The planet carrier of set 2 may rotate in the forward direction (the direction of the input shaft), but not in the reverse direction since the overrunning clutch prevents this rotation.

Gear set 1 operates in Condition 2 in Chart 3, so

$$TF_1 = Tq_{\text{out s1}}/Tq_{\text{in r1}} = -T_s/T_r, \text{ and}$$
$$TF_2 = Tq_{\text{out pc1}}/Tq_{\text{in r1}} = 1 + T_s/T_r$$

The torque of planet carrier 1 and therefore one component of the torque placed on the output shaft of the transmission is

$$Tq_{\text{out trans}} = (Tq_{\text{in gear syst}}) \cdot (1 + T_s/T_r)$$

127

The output torque of sun gear 1 is

$$Tq_{\text{out s1}} = (Tq_{\text{in gear syst}})\,(-T_s/T_r)$$

There is no torque opposing this torque except that which planet pinions 2 can provide. The speed of ring gear 2 is limited to that of the transmission output shaft, whereas there is no torque limiting the speeds of sun gear 2 and planet pinions 2 if planet carrier 2 is not Held by the overrunning clutch. The limitation on the speed of ring gear 2 causes sun gear 2 to tend to drive planet carrier 2 in the reverse direction but it is Held by the overrunning clutch. Therefore, gear set 2 is in Condition 1 in Chart 1 when sun gear 2 is considered to be the input gear.

The torque on ring gear 2 shaft, or transmission output shaft, due to the torque on sun gear 2 is

$$Tq_{\text{out trans}} = (Tq_{\text{in gear syst}})\,(-T_{s1}/T_{r1})\,(-T_{r2}/T_{s2}),\text{ or}$$
$$Tq_{\text{out trans}} = Tq_{\text{in gear syst}},\text{ since } T_{s1} = T_{s2} \text{ and } T_{r1} = T_{r2}$$

This is the second component of the transmission output-shaft torque.

The total torque of the output shaft is the sum of its two components, or

$$\text{Total } Tq_{\text{out trans}} = (Tq_{\text{in gear syst}})\,(1 + T_s/T_r) + Tq_{\text{in gear syst}}$$

The torque factor of the gear system is

$$TF_{\text{gear syst}} = (1 + T_s/T_r) + 1 = 2 + T_s/T_r = 2.45,\text{ and}$$
$$SF_{\text{gear syst}} = 1/TF = 0.408$$

since the gear system is a simple system even though one of the gear sets is in a compound system condition.

Since the maximum TF of the converter is 2.0,

$$TF_{\text{trans}} = 4.90 \text{ to } 2.45,\text{ and}$$
$$SF_{\text{trans}} = 0.408 \text{ to } 0.0$$

4. 2nd-Gear Power-On Operation (In Drive Position)

Clutch 2 and band 1 are engaged.

The power flow is from input to output shaft through gear set 1 only since the sun gear is Held by band 1 and torque cannot be transmitted through gear set 2 by ring gear 2 since the overrunning clutch overruns.

Gear set 1 is in Condition 4 in Chart 1 and the TF is $1 + T_s/T_r$. The torque of the output shaft of the transmission is

$$Tq_{\text{out trans}} = (Tq_{\text{in gear syst}})\,(1 + T_s/T_r),\text{ so}$$
$$TF_{\text{gear syst}} = 1.45,\text{ and}$$
$$SF_{\text{gear syst}} = 1/TF = 0.69$$

128

As discussed in Chapter 10, § 4,

$$TF_{\text{trans}} = 1.45 \text{ to a higher value, and}$$
$$SF_{\text{trans}} = 0.69 \text{ to a lower value}$$

5. 3rd-Gear Power-On Operation (In Drive Position)

Clutches 1 and 2 are engaged.

The engaged clutches lock ring gear 1 and sun gear 1 together and these are connected to the input shaft. Planet carrier 1 is connected to the output shaft. One of Conditions 13–18 in Chart 1 exists and the TF and SF of the gear set are 1.0.

Sun gear 2 and ring gear 2 rotate at input (and output) shaft speed, so, in effect, they are locked together. Therefore, planet carrier 2 also rotates at the same speed and the overrunning clutch overruns. The TF and SF of gear set 2, then, are 1.0 and

$$TF_{\text{trans}} = 1.0 \text{ to a higher value, and}$$
$$SF_{\text{trans}} = 1.0 \text{ to a lower value}$$

6. Reverse-Gear Power-On Operation

Clutch 1 and band 2 are engaged.

The power flow is through clutch 1, sun gear connecting member, and gear set 2. The torque on sun gear 2 is in the forward direction, and planet carrier 2 is Held by band 2. Therefore, gear set 2 is in Condition 1 in Chart 1 and the TF is $-T_r/T_s$. Then,

$$TF_{\text{gear syst}} = -2.21$$
$$SF_{\text{gear syst}} = -0.452$$
$$TF_{\text{trans}} = -4.42 \text{ to } -2.21, \text{ and}$$
$$SF_{\text{trans}} = -0.452 \text{ to } 0.0$$

7. Neutral-Gear Operation (In Neutral or Park Position)

All clutches and bands are disengaged.

The power flow is from the input shaft or output shaft (in Neutral position) to clutches 1 and 2, but no torque passes through these clutches and the TF and SF of the transmission are 0.0

8. First-Gear Power-On Operation (In First Position)

Clutch 2 and band 2 are engaged.

The clutch and band conditions are the same as when in 1st gear except that band 2 is engaged. This band Holds planet carrier 2 which is Held in the Drive or Second position by the overrunning clutch. The overrunning clutch assists in Holding the planet carrier if slippage occurs

129

between band 2 and its drum. The SF and TF, then, are the same as in 1st gear when in the Drive or Second position, or

$$TF_{\text{trans}} = 4.90 \text{ to } 2.45, \text{ and}$$
$$SF_{\text{trans}} = 0.408 \text{ to } 0.0$$

9. Second-Gear Power-On Operation (In Second Position)

The clutch and band conditions are the same as in 2nd gear when in the Drive position, so

$$TF_{\text{trans}} = 1.45 \text{ to a higher value, and}$$
$$SF_{\text{trans}} = 0.69 \text{ to a lower value}$$

10. 1st-Gear Power-Off Operation (In Drive or Second Position)

Clutch 2 is engaged.

The functions of the input and output shafts shown in Fig. 24 are exchanged. The output shaft is the "new" input shaft and it rotates in the forward direction with respect to the rotation of the crankshaft.

Gear set 1 appears to be in Condition 6 in Chart 3. If so,

$$TF_1 = TF_{\text{out s1}} = 1/(1 + T_r/T_s), \text{ and}$$
$$TF_2 = TF_{\text{out r1}} = 1/(1 + T_s/T_r)$$

The torque of sun gear 1 is

$$Tq_{\text{out s1}} = (Tq_{\text{in trans}}) \, (1/[1 + T_r/T_s])$$

This torque is applied to planet carrier 2 with a TF of $1 + T_r/T_s$ (Condition 5, Chart 1). The torque of planet carrier 2 is

$$Tq_{\text{out pc2}} = (Tq_{\text{in trans}}) \, (1/[1 + T_r/T_s]) \, (1 + T_r/T_s)$$
$$= Tq_{\text{in trans}}$$

This torque is in the forward direction, so planet carrier 2 is rotated in the forward direction by sun gear 2 and by ring gear 2. The only torque that can be provided by planet carrier 2 in opposition to the torques from sun gear 2 and ring gear 2 is that due to the forces of friction, which are small. Therefore, sun gear 2 and sun gear 1 are Free to rotate in the forward direction.

Since the sun gears are Free, one of Conditions 7–12 in Chart 1 exists in gear set 1, or Condition 6 in Chart 3 exists with substantially zero torque on both output shafts. Therefore, the torque of the "new" output shaft is substantially zero and the TF and SF of the transmission are 0.0. This means that no torque is transmitted from the wheels of the automobile to the engine and no decelerating action occurs as the result of the automobile driving the engine. However, in power-off operation, the transmission shifts to 2nd gear or Second gear when the automobile speed is above approximately 12 mph.

11. First-Gear Power-Off Operation (In First Position)

Clutch 2 and band 2 are engaged.

Since band 2 is engaged, planet carrier 2 is Held and gear set 2 is in Condition 2 of Chart 1 in which the SF is $-T_r/T_s$. The speed of sun gear 2 resulting from the rotation of the "new" input shaft is

$$S_{s2} = (S_{\text{in trans}}) (-T_r/T_s)$$

Gear set 1 operates in Condition 1 of Chart 3 and

$$SF = S_{\text{out}}/S_{\text{in 1}} = (T_s/T_r) (A-1) + A$$

where

$$A = \pm S_{\text{in 2}}/S_{\text{in 1}}$$

and $S_{\text{in 1}}$ is the speed of sun gear 1 shaft and $S_{\text{in 2}}$ is the speed of planet carrier 1 shaft. The speed of sun gear 1 shaft is the same as that of sun gear 2 shaft, and the speed of planet carrier 1 is $S_{\text{in trans}}$. Then,

$$A = (S_{\text{in trans}})/([S_{\text{in trans}}] [-T_r/T_s]) = -T_s/T_r$$

The speed of the "new" output shaft of the gear system is

$$
\begin{aligned}
S_{\text{out}} &= (S_{\text{in s1}}) ([T_s/T_r] [-T_s/T_r - 1] - T_s/T_r) \\
&= (S_{\text{in trans}}) (-T_r/T_s) ([T_s/T_r] [-T_s/T_r - 1] - T_s/T_r) \\
&= (S_{\text{in trans}}) (2 + T_s/T_r)
\end{aligned}
$$

Then,

$$SF_{\text{gear syst}} = 2.45, \text{ and}$$
$$TF_{\text{gear syst}} = 1/SF = 0.408$$

These values are the reciprocals of the power-on values when the transmission control is in the First position, or when the transmission is in 1st gear and the control is in the Drive or Second position.

Since the TF of the torque converter is 1.0 in power-off operation,

$$TF_{\text{trans}} = 0.408, \text{ and}$$
$$SF_{\text{trans}} = 2.45 \text{ to } 0.0$$

From the above, the automobile drives or tends to drive the engine when the engine is throttled to a power-off state and the transmission control is in the First position. Since the TF is less than 1.0 and the SF is greater than 1.0 when the slip of the converter is not large, the deceleration force acting on the automobile is greater than it would be with the direct drive through the transmission that occurs in 3rd gear (§ 14). Chapter 2 discusses the factors involved in this deceleration, including the torque converter factors.

131

12. Second-Gear Power-Off Operation (In Second Position)

Clutch 2 and band 1 are engaged.

Gear set 2 may be considered to be in Condition 4 of Chart 1 since sun gear 2 is Held by band 1. The torque factor is $1 + T_s/T_r$. Therefore, planet carrier 2 rotates in the forward direction and the overrunning clutch overruns. Thus (substantially), no torque is transmitted to gear set 2 and all of the torque of the "new" input shaft is applied to planet carrier 1 of gear set 1. Gear set 1 is in Condition 3 of Chart 1 and the TF is $1/(1 + T_s/T_r)$. Then,

$$TF_{\text{gear syst}} = 1/1.45 = 0.69,$$
$$SF_{\text{gear syst}} = 1/TF = 1.45,$$
$$TF_{\text{trans}} = 0.69, \text{ and}$$
$$SF_{\text{trans}} = 1.45 \text{ to a lower value}$$

Thus, the rear wheels of the automobile drive the engine when the transmission control is in the Second position and the engine is throttled to a power-off state. The "braking power" is less than when in First gear and greater than when in 3rd gear (§ 14).

13. 2nd-Gear Power-Off Operation (In Drive Position)

The clutch and band conditions are the same as in § 12 so the speed and torque factors are the same as in that section. However, the transmission automatically shifts to 3rd gear when the automobile speed is above approximately 17 mph and the throttle is closed. If this shift occurs, the situation is the one described in the following section.

14. 3rd-Gear Power-Off Operation (In Drive Position)

Clutches 1 and 2 are engaged.

Gear set 1 is in one of Conditions 13–18 in Chart 1, so sun gear 1 rotates in the forward direction at the speed of the "new" input shaft.

Ring gear 2 also rotates in the forward direction at the speed of the "new" input shaft. Therefore, ring gear 2 and sun gear 2, in effect, are locked together. Gear set 2 is in one of Conditions 13–18 in Chart 1 and rotates as a unit since the overrunning clutch overruns.

Both gear sets, then, are in one of Conditions 13–18 and their TF and SF are 1.0. Then,

$$TF_{\text{trans}} = 1.0, \text{ and}$$
$$SF_{\text{trans}} = 1.0 \text{ to a lower value}$$

Thus, the drive through the transmission gears is "direct" drive and the rear wheels of the automobile drive the engine in power-off operations. The braking effect, however, is less than when the transmission control is in the Second position and still less than when it is in the First position.

15. Reverse-Gear Power-Off Operation

The "new" input shaft rotates in the reverse direction when compared with its normal direction in the operation of the automobile. In all of the computations of the SF and TF of the gear sets, the direction of rotation of the input shaft (or input shaft 1) is considered to be the forward direction. In the following, the direction of rotation of the "new" input shaft is considered as being the reverse direction in order that the operation of the overrunning clutch is unchanged; that is, it Holds in the reverse direction and overruns in the forward direction when the forward direction is the direction of rotation of the crankshaft. Appropriate changes then must be made in the results of the computations to compensate the change in direction of the input shaft.

Clutch 1 and band 2 are engaged.

Gear set 2 is in Condition 2 in Chart 1 and its TF is $-T_s/T_r$.

Gear set 1 can transmit no torque to the "new" output shaft since ring gear 1 is Free, so its SF and TF are 0.0.

The TF of the gear system is the TF of gear set 2, so

$$TF_{\text{gear syst}} = -0.452,$$
$$SF_{\text{gear set}} = 1/TF = -2.21,$$
$$TF_{\text{trans}} = -0.452, \text{ and}$$
$$SF_{\text{trans}} = -2.21 \text{ to } 0.0$$

Since the "new" input shaft rotates in the reverse direction and the TF and SF are negative, the "new" output shaft rotates in the forward direction with respect to the direction of rotation of the crankshaft and the rear wheels of the automobile drive the engine in the forward direction in power-off operation.

16. Control System Considerations

The pattern of clutch and band engagements in the drive transmission control positions and gears, and the corresponding gear system torque factors, are shown in Table 10. The symbol, E, indicates engaged, and H or OR means that the overrunning clutch Holds or overruns. NI means that the overrunning clutch is "not involved" since band 2 Holds the planet carrier so the clutch neither Holds nor overruns, although it may assist the band in Holding the carrier. Certain of the gear shifts are used in the following to illustrate problems involved in the design of the control system.

The simplest of all of the control system problems is that presented by the power-on or power-off shifts from 2nd to Second gear, or vice versa, which involve no changes in the bands or clutches or in the relative speeds of the engine and gear system input shaft. The problem, then, is one of permitting the effective use of the up-shift control force which results from the speed of the automobile and causes the shift to 3rd gear when in the Drive position, and of preventing this use when in the Second position. Comparable in simplicity is the problem of power-on shifts from 1st to First gear or vice versa which involves only the engagement or disengagement of band 2 and the transfer of the Holding torque from the overrunning clutch to the combination of this clutch and the band or vice versa.

TABLE 10. CONTROL SYSTEM PATTERN

CP	Gear	Clutch		Band		OR Clutch		Gear Syst TF	
		1	2	1	2	P-On	P-Off	P-On	P-Off
F	First		E		E	NI	NI	2.45	0.408
S	1st		E			H	OR	2.45	0.000
	Second		E	E		OR	OR	1.45	0.690
D	1st		E			H	OR	2.45	0.000
	2nd		E	E		OR	OR	1.45	0.690
	3rd	E	E			OR	OR	1.00	1.000
R	Rev.	E			E	NI	NI	−2.21	−0.452

Other simple problems are those pertaining to power-off shifts from 3rd, 2nd or Second gear to 1st gear. In each case, disengagement of a torque-transmitting clutch or band accomplishes the shift and the status of the overrunning clutch does not change. The TF of the transmission becomes 0.0 as soon as the clutch or band is disengaged so there is no problem of near-synchronization of the engine speed with that of the "new" output shaft of the gear system.

More difficult problems are encountered in a power-on shift from 2nd to 3rd gear. In this case, band 1 is disengaged and clutch 1 is engaged during the shift. If the timing of these changes by the control system is such that both are engaged to a significant degree for a period of time, the torques of the clutch and band "fight" one another and wear occurs since the engine tends to rotate the band with the torque from clutch 1 during that time. If the timing is such that both are completely disengaged, the transmission is in 1st gear during the period of time of the disengagement. Therefore, the engine speed increases during this time, whereas it should decrease in order to provide near-synchronization of the engine speed and the speed the gear system input shaft has when clutch 1 is engaged. Thus, the control system should cause the transition to be made in such manner that band 1 and clutch 1 do not "fight" one another and so that the transmission is not in 1st gear for a time sufficient to permit an appreciable increase in the engine speed. The turbine speed is reduced by a factor of 1/1.45 during the period of the shift and the

134

engine speed needs to change by approximately this factor during the period if the acceleration of the automobile is to remain reasonably constant. Thus, the excess speed of the engine may cause a momentary surge in the speed of the automobile if the completion of the engagement of clutch 1 is accomplished too rapidly.

Problems similar to the above arise in certain power-off down-shifts. As examples, assume that when in 3rd gear the driver changes the transmission control from Drive to Second or First in order to provide greater engine "braking power." In the first case, clutch 1 and band 1 should not be engaged simultaneously for the reasons stated above. In the second instance, clutch 1 and band 2 should not be engaged simultaneously since their engagement tends to place the transmission in Reverse gear. The period of simultaneous disengagement of the clutch and band 1 or 2 should be zero or short in order that the engine does not have time to decelerate from its 3rd-gear speed prior to the engagement of band 1 or 2. In a power-off change from 3rd to Second gear at a nearly constant speed torque must be provided to increase the engine speed by a factor of approximately 1.45, and in a change from 3rd to First gear this factor is approximately 2.45. Therefore, too rapid completion of the engagement of band 1 or 2 results in fast deceleration of the automobile and stresses throughout the power train.

Power-on down-shifts from 3rd to 2nd or Second gear involve control system problems which are similar to the above except in the case of the engine-speed problem. In these down-shifts, the speed of the engine should be permitted to increase by a factor of approximately 1.45 prior to and during the engagement of band 1. Therefore, a short period of time during which clutch 1 and band 1 are disengaged, followed by a very rapid engagement of band 1 when the engine speed is 1.45 times the pre-shift gear system input shaft speed, would be desirable.

13

Powerglide Transmissions

1. General Description

Powerglide transmissions have been built in different ways in different models of the transmission. The control lever has five positions: Drive, Low, Neutral, Reverse, and Park. The throttle control serves as a transmission control device, as described in Chapter 10. The manual and automatic controls cause the transmission to shift gears in the manner described in that chapter and as indicated below.

When in the D position, the transmission shifts from 1st gear to High gear, and vice versa. When in the L, R, N, or P position, the transmission does not shift. (The terms 1st gear, Low gear, etc., and the conditions in the Park position are discussed in Chapter 2, § 4.)

Figure 25 illustrates a transmission which is the same as or similar to one of the models of the Powerglide transmissions. The gear set is of the type shown in Fig. 17. The three-element torque converter which is in the power train between the engine and the gear system has a maximum torque factor of approximately 2.0. (Other models of the transmission have more than three elements.)

§ 4 of Chapter 10 provides information pertaining to the TF and SF values as they are stated in this chapter.

2. Numbers of Gear Teeth

The numbers of teeth on the gears are assumed to be:

Sun gear 1 — 23
Planet pinions 1 — 28
Ring gear — 79
Planet pinions 2 — 18
Sun gear 2 — 28

3. 1st-Gear or Low-Gear Power-On Operation (In Drive or Low Position)

Band 1 is engaged.

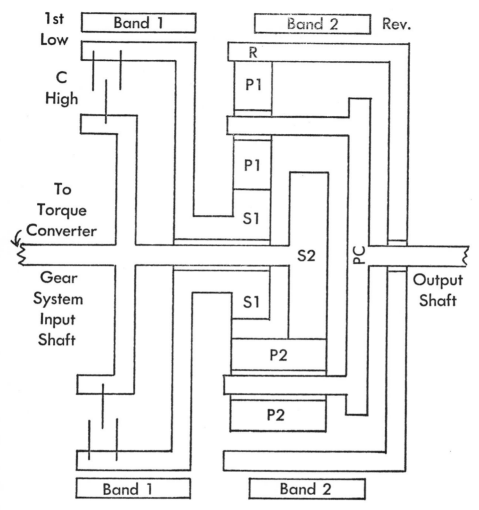

FIG. 25. Powerglide transmission. (1) 1st, High, etc., indicate band or clutch is engaged in 1st gear, etc. (2) Gear set is illustrated in Fig. 17. (3) Bands are connected to transmission case.

The gear set is in Condition 10 in Chart 4 and

$$TF_{\text{gear set}} = 1 + T_{s1}/T_{s2} = 1.82, \text{ and}$$
$$SF_{\text{gear set}} = 1/TF = 0.55$$

Since the maximum TF of the torque converter is 2.0,

$$TF_{\text{trans}} = 3.64 \text{ to } 1.82, \text{ and}$$
$$SF_{\text{trans}} = 0.55 \text{ to } 0.0$$

4. High-Gear Power-On Operation (In Drive Position)

The clutch is engaged.

The clutch locks together sun gear 1 and sun gear 2. One of Conditions 49–60 in Chart 4 exists, and the TF and SF of the gear set are 1.0. Then,

$$TF_{\text{trans}} = 1.0 \text{ to a higher value, and}$$
$$SF_{\text{trans}} = 1.0 \text{ to a lower value}$$

5. Reverse-Gear Power-On Operation

Band 2 is engaged.

The gear set is in Condition 22 in Chart 4, and

$$TF_{\text{gear set}} = 1 - T_r/T_{s2} = -1.82, \text{ and}$$
$$SF = 1/TF = -0.55$$

Then,

$$TF_{\text{trans}} = -3.64 \text{ to } -1.82, \text{ and}$$
$$SF_{\text{trans}} = -0.55 \text{ to } 0.0$$

6. Neutral-Gear Operation (In Neutral or Park Position)

The clutch and both bands are disengaged.

One of Conditions 25–36 in Chart 4 exists, and the TF and SF of the transmission are 0.0.

7. Power-Off Operation

In each of the gears the clutch and band conditions are the same in power-off and power-on operations. There is no overrunning clutch such as that in the TorqueFlite transmission. Therefore, the SF and TF of the gear system in any of the gears except Neutral are the reciprocals of those in power-on operation. However, when in the Drive position, the transmission shifts to High gear when the automobile speed is above approximately 15 mph and the throttle is closed. Since the TF of the converter is 1.0 in power-off operation, the torque factors of the transmission are the same as those of the gear system. The speed factors are the same or lower ones.

The transmission transmits torque from the wheels of the automobile to the engine in any of the gears except Neutral, and this tends to decelerate the automobile or prevent acceleration. The amounts of torque and power transmitted to the engine are discussed in Chapter 2.

8. Control System Considerations

The control system disengages band 1 and engages the clutch during shifts from 1st gear or Low gear to High gear. The opposite changes are made in the corresponding down-shifts. No change is made in the engagements in shifts from 1st to Low gear, and vice versa.

The control system should prevent the simultaneous engagement of band 1 and the clutch to significant extents since the torques of the two would "fight" one another and result in stresses on the clutch and band.

In power-on up-shifts, the period of time of simultaneous disengagement of the clutch and band 1 should be no more than a very short one since the transmission is in Neutral gear during this time. Therefore, the engine speed would increase whereas it should decrease by a factor of approximately 1/1.82. Reasonably slow completion of the engagement of the clutch is needed in order that the speed of the engine will be reduced at a rate which prevents a surge or sudden increase in the speed of the automobile. Likewise, in power-off down-shifts, the period of simultaneous disengagement should be no more than a very short one so as to prevent a decrease in engine speed, and the completion of the engagement of band 1 should be reasonably slow in order to prevent a sudden deceleration of the automobile.

In power-off up-shifts, a time of disengagement of the clutch and band sufficient to permit the speed of the engine to change by a factor of approximately 1/1.82, followed by rapid engagement of the clutch, would be desirable. Likewise, a period of time of disengagement sufficient to permit the engine speed to increase by a factor of 1.82 followed by a rapid engagement of band 1 would be desirable in power-on down-shifts.

14

Dual-Path Turbine Drive
Transmissions

1. General Description

There are five positions of the control lever: Drive, Low, Reverse, Neutral, and Park. When in the D position, the transmission shifts from 1st gear to High gear and vice versa in the manner described in Chapter 10. When in the other positions, the transmission does not shift. (The terms 1st gear, Low gear, etc., and the conditions in the Park position are discussed in Chapter 2, § 4.

Figure 26 illustrates a transmission which is the same as or similar to a Dual-Path Turbine Drive transmission. The gear set may be described as a split simple set since the sun gears have the same number of teeth and the planet pinions have the same number. The overrunning clutches are described in Chapter 2, § 6. The torque converter has a maximum *TF* of approximately 2.0.

The transmission is of unusual design in that the stator of the torque converter is used as a turbine in Reverse-gear operation and no band is used. The turbine is held stationary when the stator is used as a turbine. From Chapter 2, § 6,

$$Tq_{out} \text{ stator shaft} = Tq_{in} \text{ impeller} - 2.0\ Tq_{in} \text{ impeller}$$
$$= -Tq_{in} \text{ impeller, and}$$
$$TF = (Tq_{out} \text{ stator})/(Tq_{in} \text{ impeller}) = -1.0$$

§ 4 of Chapter 10 provides information pertaining to the *TF* and *SF* values as they are stated in this Chapter.

2. Numbers of Gear Teeth

The numbers of teeth on the gears are assumed to be:

Sun gears 1 and 2	— 46
Planet pinions 1 and 2	— 17
Ring gear	— 80

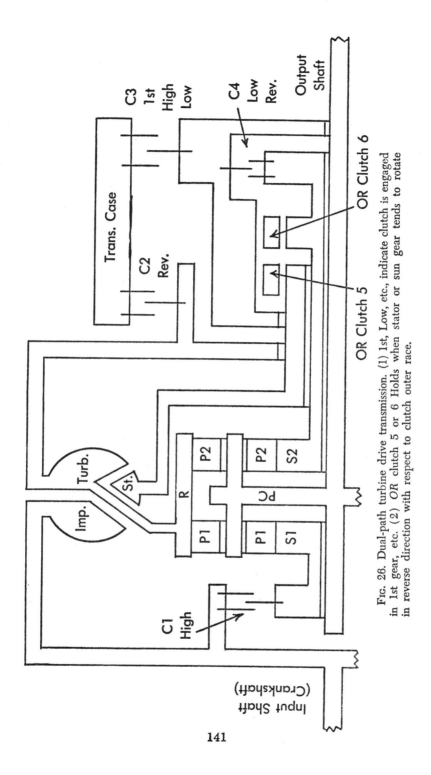

Fig. 26. Dual-path turbine drive transmission. (1) 1st, Low, etc., indicate clutch is engaged in 1st gear, etc. (2) OR clutch 5 or 6 Holds when stator or sun gear tends to rotate in reverse direction with respect to clutch outer race.

141

3. 1st-Gear Power-On Operation (In Drive Position)

Clutch 3 is engaged.

The power flow is from the input shaft to the impeller, turbine, ring gear, and pinions 1 and 2. The torque required to rotate the planet carrier and output shaft causes sun gears 1 and 2 to tend to rotate in the reverse direction, but overrunning clutch 6 and clutch 3 Hold sun gear 2 and thereby prevent rotation of sun gear 1. Therefore, the power flow continues from the ring gear to the planet carrier and output shaft. The gear set is in Condition 4 in Chart 1 and

$$TF = 1 + T_s/T_r = 1.575, \text{ or } 1.58, \text{ and}$$
$$SF = 0.635$$

Since the maximum TF of the torque converter is 2.0,

$$TF_{\text{trans}} = 3.16 \text{ to } 1.58, \text{ and}$$
$$SF_{\text{trans}} = 0.635 \text{ to } 0.0$$

4. 1st-Gear Power-Off Operation (In Drive Position)

Clutch 3 is engaged.

The power flow is from the "new" input shaft to the planet carrier. The forward torque of the carrier tends to rotate sun gears 1 and 2 in the forward direction since the speed of the carrier tends to be greater than that of the ring gear which is connected through the converter to the throttled engine. Overrunning clutch 6 overruns and the sun gears provide substantially no torque in opposition to the torque of the planet carrier. Therefore, the planet carrier and the transmission offer substantially no opposition to the rotation of the "new" input shaft and the TF and SF of the transmission are 0.0. Thus, the engine provides no braking effect in power-off operation in 1st gear. However, the transmission shifts to High gear when the automobile speed is above approximately 15 mph if the engine is throttled.

5. High-Gear Power-On Operation (In Drive Position)

Clutches 1 and 3 are engaged.

There are two power-flow paths: (1) input shaft to impeller, turbine, ring gear, pinions 1 and 2, planet carrier, and output shaft; (2) input shaft to clutch 1, sun gear 1, pinions 1, planet carrier, and output shaft.

The gear set is in Condition 5 in Chart 3 and

$$SF_1 = S_{\text{out}}/S_{\text{in s1}} = S_{\text{out}}/S_{\text{in trans}}$$
$$= (T_s + A\,T_r)/(T_s + T_r)$$

where

$$A = \pm S_{\text{in r}}/S_{\text{in s1}} = \pm S_{\text{in r}}/S_{\text{in trans}}$$

142

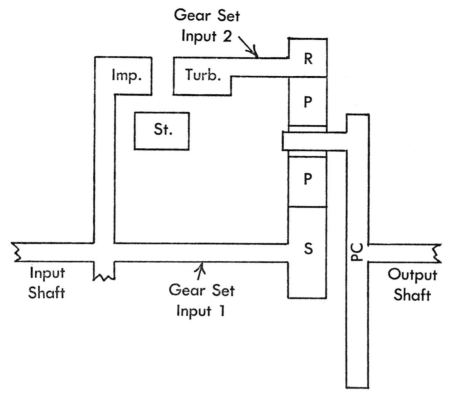

FIG. 27. Dual-path turbine drive transmission in high gear.

Figure 27 illustrates the transmission conditions when in High gear. Assume that the slip of the converter when the automobile is being driven at a constant speed of 30 mph on a level road is 3 per cent. Then,

$$SF_{\text{trans}} = S_{\text{out}}/S_{\text{in s1}} = (T_s + 0.97\,T_r)/(T_s + T_r) = 0.981$$

As indicated by the two power-flow paths, part of the power is transferred from the input shaft to the output shaft by mechanical means (gears, etc.) and part by a combination of mechanical and hydraulic means. The hydraulic means provide a dampening of the engine power pulses and otherwise provide a "smoothness" of operation. Conversely, mechanical means frequently are more efficient from the standpoint of power losses than hydraulic means. Therefore, the relative amounts of the output torque that are provided by the two paths are of interest. The ratio of these torques is

$$Tq_{\text{in imp}}/Tq_{\text{in s1}} = Tq_{\text{in}} \text{ hydr and mech path}/Tq_{\text{in}} \text{ mech path}$$

In order to determine the value of this ratio, assume any particular value of Tq_{load}. This load torque offers a torque in opposition to the input torque, $Tq_{in\ sl}$, of

$$(-Tq_{out})/(1 + T_r/T_s) = (-Tq_{out})/2.74 = -0.365\ Tq_{out}$$

since $1 + T_r/T_s$ is the TF of the input torque following path 2, and since $Tq_{load} = -Tq_{out}$.

The load torque offers a torque in opposition to the input torque, $Tq_{in\ imp}$, of

$$(-Tq_{out})/([1 + T_s/T_r]\ [TF_{conv}]) = -0.635\ Tq_{out}/TF_{conv}$$

In constant-speed High-gear cruising on level roads, TF_{conv} is approximately 1.0. Assuming that it is 1.0,

$$Tq_{in\ imp}/Tq_{in\ sl} = 0.635/0.365$$

The total input torque under the above assumption is $(0.635 + 0.365)$ (Tq_{out}), or Tq_{out}. Therefore, the flow of torque from the input to the output shaft is divided with 63.5 percent following the hydraulic-mechanical path and 36.5 per cent following the mechanical path.

Let it be assumed that the automobile acceleration and/or the road conditions are such that considerable slip exists in the converter and as a result the TF of the converter is 1.2. When Condition 5 applies (Chapter 5, § 5),

$$U = \pm Tq_{in\ 2}/Tq_{in\ 1} = T_r/T_s = 1.74$$

Therefore,

$$Tq_{in\ r}/Tq_{in\ s} = 1.74 = (Tq_{in\ imp})\ (1.2)/Tq_{in\ s},\ \ \text{or}$$
$$Tq_{in\ imp} = 1.45\ Tq_{in\ s}$$

The total Tq_{in} is $Tq_{in\ imp} + Tq_{in\ s}$, so

$$Tq_{in} = Tq_{in\ s} + 1.45\ Tq_{in\ s} = 2.45\ Tq_{in\ s},\ \ \text{and}$$
$$Tq_{in\ s} = 0.408\ Tq_{in}$$

Thus, 40.8 per cent of the input torque flows to path 2 which is the mechanical path and 59.2 percent flows to the hydraulic-mechanical path. The output torque of the transmission is

$$Tq_{out} = (Tq_{in}\ \text{path}\ 2)\ (TF_1) = (0.408\ Tq_{in})\ (2.74) = 1.118\ Tq_{in},\ \ \text{or}$$
$$Tq_{out} = (Tq_{in}\ \text{path}\ 1)\ (TF_{conv})\ (TF_2) = (0.592\ Tq_{in})\ (1.2)\ (1.575)$$
$$= 1.118\ Tq_{in},\ \ \text{and}$$
$$TF_{trans} = Tq_{out}/Tq_{in} = 1.118$$

The value of U must be $T_r/T_s = 1.74/1$, which is equal to $0.635/0.365$. Therefore, 63.5 per cent of the output torque is derived from the hydraulic-mechanical path and 36.5 percent is derived from the mechanical path as in the case when the TF of the converter is 1.0.

The divisions of torques between the two paths are the reason for designating the transmission as a Dual-Path transmission. Chapter 5, § 7 discusses Fig. 27 and the division of torques.

From the above,

$$TF_{\text{trans}} = 1.0 \text{ to a higher value, and}$$
$$SF_{\text{trans}} = 1.0 \text{ to a lower value}$$

6. High-Gear Power-Off Operation (In Drive Position)

Clutches 1 and 3 are engaged.

There are two power-flow paths: (1) "new" input shaft to planet carrier, pinions 1 and 2, ring gear, turbine, impeller, and "new" output shaft; (2) "new" input shaft to planet carrier, pinions 1 and 2, sun gears 1 and 2, and "new" output shaft. The torque on sun gear 2 is in the forward direction so overrunning clutch 6 overruns and the torque on sun gear 2 is substantially zero. The power flow to sun gear 1 continues through clutch 1 to the "new" output shaft. The gear set is in Condition 6 in Chart 3 and

$$SF_{\text{gear set}} = S_{\text{out sl}}/S_{\text{in}} = (T_s + T_r)/(T_s + B\,T_r), \text{ and}$$
$$TF_{\text{out sl}} = 1/(1 + T_r/T_s) = 0.365, \text{ and}$$
$$TF_{\text{out r}} = 1/(1 + T_s/T_r) = 0.635, \text{ where}$$
$$B = S_{\text{out r}}/S_{\text{out sl}}$$

and is always positive.

The torque placed on the "new" output shaft through path 1 is

$$Tq_{\text{out}} \text{ path } 1 = 0.635\, Tq_{\text{in}}\, TF_{\text{conv}}$$

The torque placed on this output shaft through path 2 is

$$Tq_{\text{out}} \text{ path } 2 = 0.365\, Tq_{\text{in}}$$

The converter acts as a fluid clutch when the turbine drives the impeller so its TF is substantially 1.0. Then,

$$TF_{\text{trans}} = Tq_{\text{out}}/Tq_{\text{in}} = (0.365 + 0.635)\,(Tq_{\text{in}})/Tq_{\text{in}} = 1.0, \text{ and}$$
$$SF_{\text{trans}} = 1.0 \text{ to } 0.0$$

Thus, the torque divides between the paths in the same percentage amounts as in power-on operation, and the engine serves as a brake on the speed of the automobile.

7. Low-Gear Power-On Operation (In Low Position)

Clutches 3 and 4 are engaged.

The power flow is from input shaft to impeller, turbine, ring gear, and

145

pinions 1 and 2. Forward rotation of planet pinions 2 causes sun gear 2 to tend to rotate in the reverse direction but it is Held by overrunning clutch 6 and clutches 3 and 4. The power flow, then, continues from ring gear to planet carrier to output shaft. The gear set is in Condition 4 in Chart 1 and

$$TF_{\text{trans}} = (1 + T_s/T_r)\,(TF_{\text{conv}}) = 1.58\,TF_{\text{conv}}$$
$$= 3.16 \text{ to } 1.58, \text{ and}$$
$$SF_{\text{trans}} = 0.635 \text{ to } 0.0$$

8. Low-Gear Power-Off Operation (In Low Position)

Clutches 3 and 4 are engaged.

The power flow is from the "new" input shaft to planet carrier, ring gear, turbine, impeller, and "new" output shaft. Sun gear 2 is Held by clutches 3 and 4, Condition 3 in Chart 1 exists, and

$$TF_{\text{gear set}} = 1/(1 + T_s/T_r) = 0.635, \text{ so}$$
$$TF_{\text{trans}} = 0.635, \text{ and}$$
$$SF_{\text{trans}} = 1.58 \text{ to } 0.0$$

9. Reverse-Gear Power-On Operation

Clutches 2 and 4 are engaged.

The power flow is from input shaft to impeller, to the stator which rotates in the reverse direction, to overrunning clutch 5 input section, overrunning clutch 5 output section, clutch 4, and sun gear 2. The turbine is Held by clutch 2. Overrunning clutch 5 does not hold the stator stationary since its housing (output section) can rotate with clutch 4 and sun gear 2.

Since the turbine is Held the ring gear is Held, and the gear set is in Condition 5 in Chart 1 and

$$TF_{\text{gear set}} = 1 + T_r/T_s = 2.74$$

As stated in § 1, the TF of the converter is -1.0 when the stator is used as a turbine. Then,

$$TF_{\text{trans}} = (2.74)\,(-1.0) = -2.74, \text{ and}$$
$$SF_{\text{trans}} = -0.365 \text{ to } 0.0$$

10. Reverse-Gear Power-Off Operation

Clutches 2 and 4 are engaged.

The power flow is from the "new" input shaft rotating in the reverse direction (with respect to the direction of the crankshaft) to the planet carrier, pinions 2, sun gear 2, clutch 4, and the overrunning clutch 5 housing. At this point no torque is transmitted to the stator since the housing, rotating in the reverse direction, overruns the stator shaft. Since

146

no torque is transmitted to the stator, no torque is transmitted to the "new" output shaft and the *TF* and *SF* of the transmission are 0.0.

11. Neutral-Gear Operation (In Neutral or Park Position)

All clutches are disengaged and the housing of the overrunning clutches is Free.

No torque can be transmitted through the transmission since the sun gears are Free, so the *TF* and *SF* of the transmission are 0.0.

12. Control System Considerations

The pattern of clutch engagements in the drive transmission control positions and gears, and the corresponding gear system torque factors, are shown in Table 11. The symbol, *E*, indicates engaged, and *H* or *OR* means that overrunning clutch 6 Holds or overruns. *NI* in the case of Low gear means that overrunning clutch 6 is "not involved" since clutches 3 and 4 Hold sun gear 2, although the overrunning clutch may assist clutch 4 in Holding the sun gear. When in Reverse gear, the overrunning clutch is not involved since clutch 4 couples the sun gear and overrunning clutch housing.

TABLE 11. CONTROL SYSTEM PATTERN

		Clutch				OR Clutch 6		Gear Syst TF	
CP	*Gear*	*1*	*2*	*3*	*4*	*P-On*	*P-Off*	*P-On*	*P-Off*
D	1st			E		H	OR	1.58	0.000
	High	E		E		OR	OR	1.00	1.000
L	Low			E	E	NI	NI	1.58	0.635
R	Rev.		E		E	NI	NI	2.74	0.000

The problem of the control system in power-on shifts from 1st to High gear is one of engaging clutch 1 slowly enough to prevent the automobile from surging forward as the result of an engine speed which needs to be reduced reasonably slowly by a factor of approximately 1/1.58 during the period of the shift if the automobile acceleration is to remain substantially constant. No problem is encountered in the timing of the transfer of torque from the overrunning clutch to clutch 1 since the overrunning clutch is completely "engaged" when the torque on the sun gear is in the reverse direction, and it completely and immediately "disengages" when this torque changes to a forward direction.

Power-on down-shifts from High to 1st gear involve changes in the speed of rotation of the sun gears from the forward-direction speed of the crankshaft to a substantially zero speed in the reverse direction and then to a zero speed. As in power-on up-shifts, the overrunning clutch

147

is completely "disengaged" during the reduction of the speed to zero and it becomes completely "engaged" immediately after the speed changes to a reverse direction. If clutch 1 is disengaged very rapidly, the TF of the transmission is zero until the engine begins to provide power to the transmission, that is, until the engine speed is approximately 1.58 times its value before the shift started. Therefore, no deceleration of the automobile results from the braking effect caused by a too-low engine speed. A reasonably slow disengagement of clutch 1 permits the engine speed to increase during the disengagement and prevents the TF from becoming zero.

No engine-speed problem exists in power-off shifts from High to 1st gear since the power-off 1st-gear TF is zero. In power-off shifts from 1st gear to High gear, the engine speed needs to be changed to one near that of the output shaft. This adjustment is a minor one if the up-shift occurs at the minimum up-shift speed of approximately 15 mph since the idling speed of the engine corresponds reasonably closely with the engine speed at 15 mph when in High gear. If the power-off up-shift occurs at a higher automobile speed as the result of a "lift-foot" change in the accelerator position, the deceleration of the engine during the shift will result in a proper engine speed adjustment if the control system causes an engagement of clutch 1 which is not too rapid, but not so slow as to permit a decrease in the engine speed by a factor of less than approximately 1/1.58.

Power-on shifts from 1st to Low gear and vice versa present no problems since the change that is made affects only the means that are used in Holding sun gear 2. Power-off shifts from Low to 1st gear present no problems since the TF of the transmission becomes zero during the shift.

In power-off shifts from 1st to Low gear, an engine speed adjustment may be needed so the control system should cause the engagement of clutch 4 to be made at a rate which prevents a sudden deceleration of the automobile and the resultant stresses in the power train. These shifts normally occur at low automobile speeds so the engine-speed adjustment is not a large one. The corresponding shift from High to Low gear, however, involves an engine speed change by a factor of approximately 1.58. Therefore, completion of the engagement of clutch 4 should be made reasonably slowly. The control system should not permit simultaneous engagement of clutches 1 and 4 to significant degrees since clutch 1 drives the sun gears and clutch 4 (with clutch 3) Holds these gears.

15

Super Turbine "300" Transmissions

1. General Description

There are five positions of the transmission control lever: Drive, Low, Reverse, Neutral, and Park. When in the D position the transmission shifts from 1st gear to High gear and vice versa in the manner described in Chapter 10. When in the other positions it does not shift. The terms 1st gear, Low gear, etc., and the conditions in the Park position are discussed in Chapter 2, § 4.

Figure 28 illustrates a transmission which is the same as or similar to a Super Turbine "300" transmission. The gear set is of the type shown in Fig. 17. The gear system is very similar to that shown in Fig. 25, a principal difference being in the use of clutch 2 in place of band 2.

The transmission control system includes an electrical switch which is associated with the throttle control. The opening and closing of this switch result in changes in the angle of the torque converter stator vanes and in the torque factor of the three-element converter. The vanes are in the high-angle position and the TF varies from 1.0 to 2.7 when the throttle is in the idling speed position and when it is more than three-fourths open. Other positions of the throttle cause the vanes to be in the low-angle position and the TF varies from 1.0 to 2.0.

When the throttle is in the idling speed position and the automobile is stationary, the difference between the speeds of the impeller and the turbine is approximately 500 rpm. The combination of the low engine speed and high-angle stator vane position results in less torque being applied to the output shaft of the converter and therefore less tendency of the automobile to move (or creep) than would exist if the vanes were in the low-angle position. The variable-stator converter is discussed in Chapter 2, § 6.

§ 4 of Chapter 10 provides information pertaining to the TF and SF values as they are stated in this chapter.

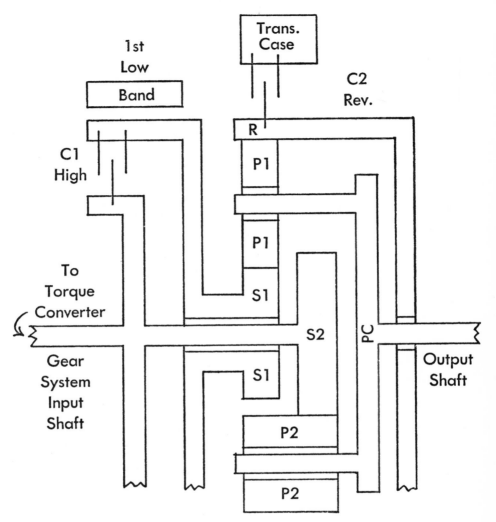

Fig. 28. Super turbine "300" transmission. (1) 1st, High, etc., indicate the band or clutch is engaged in 1st gear, etc. (2) Gear set is illustrated in Fig. 17. (3) Band is connected to transmission case.

2. Numbers of Gear Teeth

The numbers of teeth on the gears are assumed to be:

Sun gear 1 — 26
Planet pinions 1 — 34
Ring gear — 94
Sun gear 2 — 34
Planet pinions 2 — 26

150

3. 1st-Gear or Low-Gear Power-On Operation (In Drive or Low Position)

The band is engaged.

Sun gear 1 is Held, the ring gear is Free, the gear set is in Condition 10 in Chart 4, and

$$TF_{\text{gear set}} = 1 + T_{s1}/T_{s2} = 1.765, \text{ and}$$
$$SF_{\text{gear set}} = 0.566$$

Since the maximum TF of the converter is 2.7,

$$TF_{\text{trans}} = 4.77 \text{ to } 1.765, \text{ and}$$
$$SF_{\text{trans}} = 0.566 \text{ to } 0.0$$

4. High-Gear Power-On Operation (In Drive Position)

Clutch 1 is engaged.

One of Conditions 49–60 in Chart 4 exists and the TF and SF of the gear set are 1.0. Then,

$$TF_{\text{trans}} = 1.0 \text{ to a higher value, and}$$
$$SF_{\text{trans}} = 1.0 \text{ to a lower value}$$

5. Neutral-Gear Operation (In Neutral or Park Position)

The band and clutches are disengaged.

One of Conditions 25–36 in Chart 4 exists and the TF and SF of the transmission are 0.0.

6. Reverse-Gear Power-On Operation

Clutch 2 is engaged.

The gear set is in Condition 22 in Chart 4 and

$$TF_{\text{gear set}} = 1 - T_r/T_{s2} = -1.765,$$
$$SF_{\text{gear set}} = -0.566,$$
$$TF_{\text{trans}} = -4.77 \text{ to } -1.765, \text{ and}$$
$$SF_{\text{trans}} = -0.566 \text{ to } 0.0$$

7. Power-Off Operation

In any of the transmission control positions the band and clutch conditions are the same when operating with power on and power off, and there is no overrunning clutch in the gear system. Therefore, except in the Neutral or Park position, the TF and SF of the gear system in power-off operations are the reciprocals of those in power-on operations. However, the transmission will not remain in 1st gear in power-off operation when the automobile speed is above approximately 15 miles per hour.

151

The power-off TF and SF values of the transmission are:

1st and Low gears	—	$TF = 0.566$, $SF = 1.765$ to 0.0;
Reverse gear	—	$TF = -0.566$, $SF = -1.765$ to 0.0;
High gear	—	$TF = 1.0$, $SF = 1.0$ to a lower value.

8. Control System Considerations

The pattern of clutch and band engagements in forward-drive control positions and gears is the same as in § 8 of Chapter 13, which pertains to the transmission shown in Fig. 25. Therefore, the control system problems are the same as those in that Chapter when the speed factor of 1.765 is substituted for 1.82 and clutch 1 in Fig. 28 is considered to be the clutch in the discussion in Chapter 13.

16

Super Turbine "400" Transmissions

1. General Description

The five positions of the transmission control lever are Drive, Low, Reverse, Neutral, and Park. When in the D position, the transmission shifts from 1st to 2nd to 3rd gear and down-shifts in the manner described in Chapter 10. It does not shift out of Low gear when in the L position. However, it shifts to Second gear when the transmission control lever is moved from the D position to the L position while in 3rd gear. It does not up-shift from Second gear while in the L position, but it down-shifts to and remains in Low gear if the automobile speed decreases to a value of approximately 20 mph. The terms 1st gear, Low gear, etc., and the conditions in the Park position are discussed in Chapter 2, § 4. Engine intake-manifold vacuum and the throttle control are used in controlling the operations of the transmission.

Figure 29 illustrates a transmission which is the same as or similar to a Super Turbine "400" transmission. It employs overrunning clutches which are described in Chapter 2, § 6. The torque converter indicated in the illustration has a maximum TF of approximately 2.1. Some models of the transmission have a variable-pitch stator.

Section 4 of Chapter 10 provides information pertaining to the TF and SF values as they are stated in this Chapter.

2. Numbers of Gear Teeth

The numbers of teeth on the gears are assumed to be:

Sun gears 1 and 2	— 26
Planet pinions 1 and 2	— 14
Ring gears 1 and 2	— 54

3. 1st-Gear Power-On Operation (In Drive Position)

Clutch 1 is engaged.

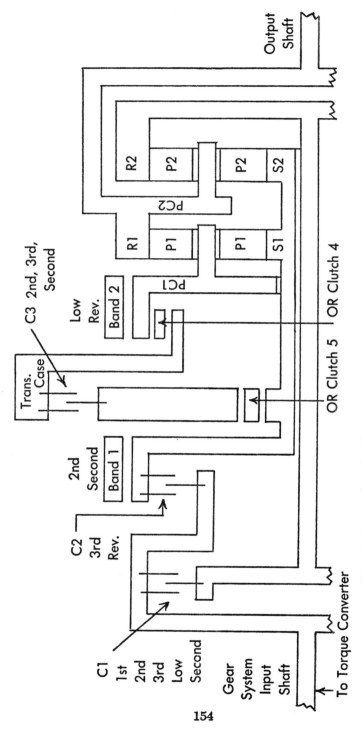

FIG. 29. Super turbine "400" transmission. (1) 1st, Low, etc., indicate the band or clutch is engaged in 1st gear, etc. (2) OR clutch 4 or 5 Holds when planet carrier 1 or sun gear tends to rotate in reverse direction. (3) Bands are connected to transmission case.

154

Power flows from the gear-system input shaft to clutch 1 and ring gear 2. Gear set 2 is in Condition 2 in Chart 3 and

$$TF_1 = Tq_{\text{out s2}}/Tq_{\text{in r2}} = -T_s/T_r, \text{ and}$$
$$TF_2 = Tq_{\text{out pc2}}/Tq_{\text{in r2}} = 1 + T_s/T_r$$

The torque applied to sun gear 2 (in the reverse direction) in turn is applied to sun gear 1, which tends to rotate planet carrier 1 in the reverse direction but it is prevented from doing so by overrunning clutch 4. The torque on sun gear 2 then produces a torque of $-T_r/T_s$ on ring gear 1 and the output shaft per lb-ft of torque applied to sun gear 2. The component of the torque on the output shaft due to this torque is $(-T_s/T_r)(-T_r/T_s)$, or 1.0 lb-ft per lb-ft of gear system input shaft torque.

The total torque on the output shaft is the sum of its two components, the first of which is $(Tq_{\text{in}} \text{ gear syst})(TF_2)$, so

$$Tq_{\text{total}} = 1.0 + (1 + T_s/T_r) = 2.48 \text{ lb-ft per lb-ft}$$

of gear-system input shaft torque. Then,

$$TF_{\text{gear syst}} = 2.48,$$
$$SF_{\text{gear syst}} = 0.403,$$
$$TF_{\text{trans}} = 5.21 \text{ to } 2.48, \text{ and}$$
$$SF_{\text{trans}} = 0.403 \text{ to } 0.0$$

4. 1st-Gear Power-Off Operation (In Drive Position)

The clutch and band conditions are the same as in § 3. The torque on planet carrier 1, however, is in the forward direction so the planet carrier overrunning clutch overruns. Sun gear 2 is Free to rotate in the forward direction. Therefore, pinions 1 and 2 are Free and no torque is provided in opposition to the rotation of the "new" input shaft, and no rotation of the "new" output shaft results from the rotation of the "new" input shaft. Then, the TF and SF of the transmission are 0.0.

The transmission shifts to 2nd gear when the automobile speed is above a few miles per hour and the throttle is closed.

5. Low-Gear Power-On Operation (In Low Position)

Clutch 1 and band 2 are engaged.

The band and clutch conditions are the same as when in 1st gear except that band 2 is engaged. Thus, band 2 may assist the overrunning clutch in Holding planet carrier 1 but the operation of the transmission otherwise is not changed. The TF and SF of the transmission, then, are the same as those in § 3.

6. Low-Gear Power-Off Operation (In Low Position)

Clutch 1 and band 2 are engaged.

The use of band 2 prevents rotation of planet carrier 1. The power flow paths, then, are the same as in power-on operation, but in the opposite directions. The TF and SF of the gear system in power-off operation are the reciprocals of those in power-on operation, so

$$TF_{\text{gear syst}} = 0.403,$$
$$SF_{\text{gear syst}} = 2.48,$$
$$TF_{\text{trans}} = 0.403, \text{ and}$$
$$SF_{\text{trans}} = 2.48 \text{ to } 0.0$$

7. 2nd-Gear or Second-Gear Power-On Operation (In Drive or Low Position)

Clutches 1 and 3 and band 1 are engaged.

The power flow is from the input shaft to clutch 1, ring gear 2, planet carrier 2, and the output shaft, since sun gears 1 and 2 are Held by band 1 and planet carrier 1 overruns. Gear set 2 is in Condition 4 in Chart 1, and

$$TF = 1 + T_s/T_r = 1.48,$$
$$SF = 0.675,$$
$$TF_{\text{trans}} = 1.48 \text{ to a higher value, and}$$
$$SF_{\text{trans}} = 0.675 \text{ to a lower value}$$

8. 2nd-Gear or Second-Gear Power-Off Operation (In Drive or Low Position)

Clutches 1 and 3 and band 1 are engaged.

The planet carrier 1 overrunning clutch overruns when power-on or power-off conditions exist. Therefore, the TF and SF of the gear system are the reciprocals of the values with power on, and

$$TF_{\text{trans}} = 0.675, \text{ and}$$
$$SF_{\text{trans}} = 1.48 \text{ to a lower value}$$

9. 3rd-Gear Power-On Operation (In Drive Position)

All clutches are engaged.

Clutches 1 and 2 lock together ring gear 2 and sun gears 1 and 2. Therefore, the gear sets rotate as a unit and overrunning clutches 4 and 5 overrun. One of Conditions 13–18 in Chart 1 exists, and

$$TF_{\text{trans}} = 1.0 \text{ to a higher value, and}$$
$$SF_{\text{trans}} = 1.0 \text{ to a lower value}$$

10. 3rd-Gear Power-Off Operation (In Drive Position)

All clutches are engaged.

156

The conditions in the gear sets are the same as in power-on operation, so the TF and SF of the gear system are 1.0. Then,

$$TF_{trans} = 1.0, \text{ and}$$
$$SF_{trans} = 1.0 \text{ to a lower value}$$

11. Reverse-Gear Power-On Operation

Clutch 2 and band 2 are engaged.

The power flow is from the input shaft to clutch 2, sun gear 1, planet carrier 1, ring gear 1, and the output shaft. Gear set 1 is in Condition 1 in Chart 1 since planet carrier 1 is Held. Ring gear 2 is Free. Then,

$$TF_{gear\ syst} = -T_r/T_s = -2.08,$$
$$SF_{gear\ syst} = -0.48,$$
$$TF_{trans} = -4.37 \text{ to } -2.08, \text{ and}$$
$$SF_{trans} = -0.48 \text{ to } 0.0$$

12. Reverse-Gear Power-Off Operation

All clutch and band conditions are the same as in power-on operation and the overrunning clutches are not involved in either case, so the gear system TF and SF are the reciprocals of those in power-on operation, and

$$TF_{trans} = -0.48, \text{ and}$$
$$SF_{trans} = -2.08 \text{ to } 0.0$$

13. Neutral-Gear Operation (In Neutral or Park Position)

No clutches or bands are engaged.

When clutches 1 and 2 are disengaged, there is no connection between the gear system input shaft and the remainder of the transmission. Therefore, the TF and SF of the transmission are 0.0.

14. Control System Considerations

The pattern of clutch and band engagements in the drive transmission control positions and gears, and the corresponding gear system torque factors, are shown in Table 12. The symbol, E, indicates engaged, and H or OR means that the overrunning clutch Holds or overruns. NI means that the overrunning clutch is "not involved." In the case of overrunning clutch 4, the clutch is not involved since band 2 Holds planet carrier 1, although the clutch may assist the band in Holding the carrier. Overrunning clutch 5 is not involved when clutch 3 is not engaged, and it is not involved when band 1 is fully engaged except in the sense that it may assist band 1 in Holding the sun gears. However, the control system does not engage band 1 until clutch 3 is fully engaged and disengages it before engaging clutch 2. Thus, overrunning clutch 5 Holds during any period of time the torque on the sun gears is in the reverse direction,

157

clutch 3 is engaged, and band 1 is not engaged. Therefore, it is listed as *PI*, meaning "partially involved." Certain of the gear shifts in Table 12 are used to illustrate problems involved in the design of the control system.

TABLE 12. CONTROL SYSTEM PATTERN

		Clutch			Band		OR Clutch 4		OR Clutch 5		Gear Syst TF	
CP	Gear	1	2	3	1	2	P-On	P-Off	P-On	P-Off	P-On	P-Off
D	1st	E					H	OR	NI	NI	2.48	0.000
	2nd	E		E	E		OR	OR	PI	NI	1.48	0.675
	3rd	E	E	E			OR	OR	OR	OR	1.00	1.000
L	Low	E				E	NI	NI	NI	NI	2.48	0.403
	Sec.	E		E	E		OR	OR	PI	NI	1.48	0.675
R	Rev.		E			E	NI	NI	NI	NI	−2.08	−0.480

Engagement of clutch 3 by the control system at a rate which reduces the engine speed without causing a surge in the speed of the automobile, followed by engagement of band 1, will produce a smooth power-on shift from 1st to 2nd gear when in the Drive position. Likewise, disengagement of band 1 followed by engagement of clutch 2 at a reasonably slow rate will produce a smooth shift from 2nd to 3rd gear. In power-on shifts from 3rd to 2nd gear a reasonably slow disengagement of clutch 2 will permit the engine speed to increase during the period of the transfer of sun gear torque from this clutch to overrunning clutch 5 and clutch 3. This step in the shifting process may be followed by engagement of band 1 with no change other than in the means of Holding the sun gears. Power-on shifts from 2nd to 1st gear present no automobile deceleration problems due to the automobile driving the engine since the power-off *TF* becomes zero when band 1 is disengaged. The release of band 1 followed by a reasonably slow disengagement of clutch 3 produces a shift with no interruption in the torque transmission to the wheels which would occur if these were disengaged instantaneously since a short period of time is needed to permit a change in the speed of the engine by a factor of approximately 2.48/1.48.

Power-off shifts from 1st to 2nd gear in which the engine is throttled throughout the period beginning with the starting of the automobile present no major problems of engine speed adjustment. The idling speed of the engine causes a power-on condition to exist when the automobile is starting and the shift to 2nd gear occurs at an automobile speed corresponding reasonably closely with the engine idling speed. Therefore, the above power-on timing of the engagements of clutch 3 and band 1, with reasonably slow engagement of band 1, will produce smooth shifts in which the automobile acceleration does not change to objectionable extents. Likewise, in "lift-foot" power-off shifts, in which the throttle is partially or fully closed following an acceleration of the automobile in 1st gear, a slow engagement of band 1 permits the engine speed to decrease to the required value while the TF is zero, or nearly so. However, an engagement which is too slow may permit the engine speed to decrease to a speed below that needed in 2nd gear and band 1 then would need to transmit torque to increase this speed, which would cause a deceleration of the automobile. In a similar manner, the power-on shift timing will produce a smooth power-off shift from 2nd to 3rd gear if the engagement of clutch 2 is not delayed too much.

Power-off changes of the control lever from the Drive to the Low position are made to cause the engine to provide greater "braking power." In shifts from 3rd gear, disengagement of clutch 2 followed by an immediate partial engagement of band 1 and a slow completion of that engagement would prevent the TF of the transmission becoming zero for an appreciable time, and cause the engine speed to increase at a rate which prevents a fast deceleration of the automobile during the shift. Likewise, in power-off shifts from Second to Low gear, immediate partial engagement of band 2 after disengagement of band 1 and slow completion of the engagement prevents a decrease of the TF to zero and causes a slow increase in the speed of the engine, thus preventing a fast deceleration of the automobile during the shift.

17

Dynaflow Transmissions

1. General Description

The five positions of the transmission control lever are Drive, Low, Reverse, Neutral, and Park.

The Dynaflow transmissions, and corresponding models of the Power-glide transmissions, are automatic transmissions which differ from those discussed in other chapters in that they do not shift gears while operating in any of the transmission control positions. However, when in the Low, Drive, and Reverse positions, they change gear ratios when the automobile changes from low to high speeds and vice versa, or from heavy to light load conditions and vice versa. This is accomplished without the use of bands or clutches (excepting overrunning clutches), as described in § 4.

Figure 30 illustrates a transmission which is the same as or similar to one model of the Dynaflow transmission. It employs three overrunning clutches which are described in Chapter 2, § 6. The torque converter has five elements, and stator 2 has variable-pitch vanes. The vanes are in the high-pitch position when the throttle is fully opened and in the low-pitch position when the throttle is in other positions.

§ 4 of Chapter 10 provides information pertaining to the *TF* and *SF* values as they are stated in this chapter.

2. Torque Converter and Associated Gear Set

The torque converter shown in Fig. 30 has one impeller, two turbines, and two stators. The input shaft of the converter (the engine crankshaft) is connected to the impeller as usual, but two output shafts, connected to turbines 1 and 2, are used. These shafts are connected permanently to the ring gear and planet carrier of gear set 1, which is included within the converter structure. The sun gear of this gear set is Held by an over-running clutch when it tends to rotate in the reverse direction and the transmission is in any control position except the Low position. It is Held by a band when the control lever is in the Low position. The stators are

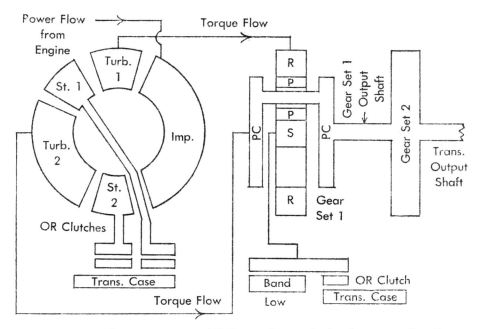

Fig. 30. Dynaflow transmission. (1) Low indicates the band is engaged in Low gear. (2) OR clutches Hold when stators or sun gear tend to rotate in reverse direction. (3) Gear set 2 is illustrated in Fig. 25. (4) Band is connected to transmission case.

Held by overrunning clutches when they tend to rotate in the reverse direction. As stated above, stator 2 has variable-pitch vanes. The output shaft of gear set 1 is the input shaft of gear set 2, which is one such as that shown in Fig. 17. This gear set is connected to bands, a clutch, etc., as shown in Fig. 25, but in this case the gear portion of the transmission in that illustration does not shift gears automatically.

When the torque converter is in a stalled condition (when the speed of the output shaft of gear set 1 is zero), turbine 1 provides the major part of the torque of the output shaft of gear set 1. When the ratio of the speeds of the impeller and this output shaft is less than approximately 1.6, turbine 2 provides all of the torque on the output shaft of gear set 1. The torque factors of the converter and gear set 1 are:

$TF = 3.4$ when the converter is stalled and stator 2 vanes are in the high-pitch position,

$TF = 3.1$ when the converter is stalled and the stator vanes are in the low-pitch position,

$TF = 1.0$ to (substantially) one or the other of the above values in power-on operation when the converter is not stalled,

$TF = 1.0$ in power-off operation.

161

3. Numbers of Gear Teeth

The numbers of teeth on the gears of gear set 2 are assumed to be those stated in Chapter 13. The numbers on gear set 1 are assumed to be:

Sun gear — 47
Planet pinions — 16
Ring gear — 79

4. Drive-Gear Power-On Operation (In Drive Position)

The clutch of gear set 2 (Fig. 25) is engaged and the bands are disengaged. The *SF* and *TF* of this gear set are 1.0 in power-on or power-off operation as long as the transmission control is in the Drive position.

Assume that the output shaft of gear set 1 is stalled or is rotating with a speed which is very low when compared with that of the impeller. The torque acting on the sun gear is the sum of the torques that are placed on it by turbines 1 and 2. The torque that is produced on it by turbine 1 is

$$Tq_{\text{in s}} \text{ from turbine } 1 = (Tq_{\text{out turb 1}})(-T_s/T_r)$$
$$= -0.60\, Tq_{\text{out turb 1}}$$

The torque on the planet carrier that is produced by turbine 1 (Condition 4, Chart 1, or Condition 2, Chart 3) is

$$Tq_{\text{in pc}} \text{ from turbine } 1 = (Tq_{\text{out turb 1}})(1 + T_s/T_r)$$
$$= 1.6\, Tq_{\text{out turb 1}}$$

This torque is in the forward direction, and in producing it the ring gear tends to move the planet pinions in the forward direction with respect to their shafts. This movement causes such clearances as exist between the pinions and their shafts to be on the forward sides of the shafts unless turbine 2 produces a torque on the shafts which prevents these positions of the clearances. Since the output shaft of gear set 1 is stalled, or nearly so, turbine 2 can provide no such torque and the clearances are on the forward sides of the shafts. Therefore, turbine 2 produces no torque on the sun gear, and all of its torque is applied to the output shaft of gear set 1.

The total torque on the sun gear, then, is $-0.60\, Tq_{\text{out turb 1}}$. The sun gear is Held by the overrunning clutch and gear set 1 is in Condition 4 in Chart 1 insofar as the input from turbine 1 is concerned.

The components of the torque on the output shaft of gear set 1 when the sun gear is Held are

$$Tq_{\text{out 1}} = (1.0)(Tq_{\text{out turb 2}}), \text{ and}$$
$$Tq_{\text{out 2}} = 1.6\, Tq_{\text{out turb 1}}$$

If it is assumed that the output torques of turbines 1 and 2 are equal, the total output torque of gear set 1 will be

$$Tq_{\text{out total}} = 2.6 \, Tq_{\text{out turb 1}}$$

The percentages of the total output torque provided by each of the turbines will be

By turbine 1 = (100%) (1.6/2.6) = 61.5%, and
By turbine 2 = (100%) (1/2.6) = 38.5%

of output torque of gear set 1.

The component of the velocity of the converter fluid in planes which are perpendciular to the axis of the converter may be greatest at the points of entry into turbine 1. This velocity, together with the greater torque arm length of turbine 1 when compared with that of turbine 2, may make the output torque from turbine 1 greater than that of turbine 2. If it is greater, the percentage of the output torque of gear set 1 that is provided by turbine 1 is greater than the value of 61.5 per cent above when the output shaft of gear set 1 is stalled. Conversely, if the torque from turbine 1 is less than that from turbine 2, less than 61.5 per cent of the torque is provided by turbine 1.

Assume that the speed of the impeller remains constant, the load provided by the automobile has considerable inertia, and the output torque of gear set 1 is sufficient to increase the speed of the input shaft of the load (the driveshaft) to a speed nearly equal to that of the impeller. As the speeds of the turbines increase while the sun gear is Held, the ratio of their speeds is

$$S_{\text{turb 2}}/S_{\text{turb 1}} = SF \text{ in Condition 4, Chart 1}$$
$$= 1/(1 + T_s/T_r) = 0.625, \text{ or}$$
$$S_{\text{turb 1}}/S_{\text{turb 2}} = 1.6$$

As the speed of turbine 1 increases, its slip decreases, and the output torque of the turbine decreases. The same occurs with respect to turbine 2, but any given percentage of decrease in torques occurs first at turbine 1 since its speed is 1.6 times that of turbine 2 as long as the sun gear is Held. When the speed of turbine 1 is slightly less than the speed of the impeller, its slip is very small and its output torque to the gear set is almost zero. At this point, turbine 2 begins to drive the sun gear in the forward direction. When the sun gear is not Held by the overrunning clutch but when its speed is still zero, the speeds of turbines 1 and 2 are such that

$$S_s = 0.0 = (S_{\text{turb 1}}) \, (-T_r/T_s) + (S_{\text{turb 2}}) \, (1 + T_r/T_s), \text{ or}$$
$$S_{\text{turb 2}}/S_{\text{turb 1}} = 1/(1 + T_s/T_r) = 0.625, \text{ and}$$
$$S_{\text{turb 1}}/S_{\text{turb 2}} = 1.6$$

163

When the ratio of the speeds of turbines 2 and 1 is greater than 0.625, the sun gear rotates freely in the forward direction with a speed of S_s as determined by the equation above, disregarding the 0.0 value.

Assume that the impeller speed still remains constant, and that the speeds of turbines 1 and 2 are substantially stabilized at values such that $S_{turb\ 2}/S_{turb\ 1}$ is greater than 0.625. The sun gear, then, rotates in the forward direction with a speed of S_s. If the speed of turbine 1 increases slightly, the speed of the sun gear decreases. When the speed of turbine 1 increases, the torque produced on it is less since the slip is less. Therefore, its speed decreases and the speed of the sun gear increases. If the speed of turbine 1 decreases to a value below the substantially stabilized value above, the sun gear speed increases. The slower turbine 1 speed causes more torque to be produced on it because of the increase in slip. Therefore, its speed increases. Thus, the speed of turbine 1 becomes substantially stable with a value such that the torque produced on it is equal to the torque required by the forces of friction which tend to decrease the speed of the turbine. Therefore, turbine 1 provides no torque to the load, and turbine 2 provides all of the output torque of gear set 1 when the ratio of the speeds of turbines 2 and 1 is greater than 0.625. The torque factor of the converter and gear set 1 is the torque factor of turbine 2. This factor has a value which varies from (substantially) 1.0 to a higher value, depending on the slip of turbine 2.

The conditions which cause the various stages of the change in the *TF* from a high value to substantially 1.0 are of interest, particularly when comparing the operations in the Drive and Low positions of the control lever.

Let it be assumed that an automobile is starting on a level road when the control lever is in the Drive position, and that the maximum possible engine speed is 1600 rpm when the converter is stalled. Assume, further, that the driver controls the throttle so that the engine speed remains constant at its maximum-torque speed, say 2500 rpm, or fully opens the throttle when the engine speed is substantially less than this value. The throttle must be fully opened if the engine is to provide the maximum possible torque at the instant the automobile starts to move. Therefore, the vanes of stator 2 are in the high-angle position and the *TF* of the transmission (§ 2) is 3.4. Soon after the automobile starts moving, the driver moves the throttle to a less than fully opened position in order to maintain the engine speed constant at 2500 rpm. The stator vanes then move to the low-angle position and the *TF* of the transmission is reduced accordingly. The *TF* also is reduced because of the reductions in the slips of the turbines. The *TF* of gear set 1 for the torque of turbine 1 is 1.6, whereas that for the torque from turbine 2 is 1.0. The speed of turbine 2 is directly proportional to the automobile speed since its output shaft rotates at the same speed as that the driveshaft. The speed of turbine 1 is 1.6 times that of turbine 2 until the decrease in slip of turbine 1 causes its torque to decrease to the point that the sun gear rotates in the forward direction. At some point during the acceleration of turbine 1, the fluid leaving the turbine begins to produce a forward-direction torque on stator 1. Therefore, it rotates in the forward direction and ceases performing the functions of a stator. As the automobile acceleration continues, the speed of turbine 2 increases, its slip decreases, and its TF decreases. When the slip reaches a particular value, stator 2 starts to rotate. At this point, and with lesser slips, the entire converter operates as a fluid clutch with a TF of substantially 1.0. Since turbine 1 provides no torque to the output shaft of gear set 1 when the sun gear rotates, the (one) input shaft and the output shaft of the gear set are locked together by the planet carrier and the TF of the gear set is 1.0. The speed of the automobile becomes stabilized, assuming that the engine speed and road conditions remain constant, at a value such that the slip of turbine 2 is that which is required to provide the torque needed by the automobile at its speed.

Summarizing, in power-on operation in the Drive position,

$$TF_{\text{trans}} = 3.4 \text{ to } 1.0, \text{ and}$$
$$SF_{\text{trans}} = 1.0 \text{ to } 0.0$$

5. Drive-Gear Power-Off Operation (In Drive Position)

As stated in § 4, the TF and SF of gear set 2 are 1.0.

Let it be assumed that a component of the forward-direction torque of the "new" input shaft of gear set 1 is transmitted to the ring gear and turbine 1 with the appropriate torque factor, which is $1/(1 + T_s/T_r)$. When the planet carrier transmits this component of the input torque to the ring gear, it also transmits torque to the sun gear (Chapter 5, § 2), and this torque is

$$Tq_{\text{out s}} = (Tq_{\text{in}} \text{ component}) (1/[1 + T_r/T_s])$$

which is in the forward direction. Therefore, the overrunning clutch would overrun, the input torque component would be substantially zero, and the torque transmitted to turbine 1 would be substantially zero.

Turbines 1 and 2 may act as impellers and the impeller acts as a turbine (or more as a fluid clutch runner) in power-off operation. Since the torque on turbine 1 from the gear set is substantially zero, there is substantially no torque to change the speed of turbine 1 to a value above or below the radial-plane component of the speed of the converter fluid at the points of passage of the fluid through turbine 1. This fluid speed component is approximately equal to the speed of turbine 2 since the stator overrunning clutches overrun in power-off operation. Therefore, the speeds of turbines 1 and 2 remain approximately equal at all times and turbine 2 is the only turbine that acts effectively as an impeller.

165

All of the torque of the "new" input shaft, then, is transferred to turbine 2 with a torque factor of 1.0. The torque on the impeller is that of the "new" input shaft, and this torque tends to drive the engine in the forward direction. The extent to which this provides braking action on the speed of the automobile is discussed in Chapter 2, § 6. The torque and speed factors are

$$TF_{trans} = 1.0, \text{ and}$$
$$SF_{trans} = 1.0 \text{ to } 0.0$$

6. Low-Gear Power-On Operation (In Low Position)

The conditions in the converter and gear set 1 are the same as in power-on operation in the Drive position except that the sun gear is Held by the band. Gear set 2 is in a condition which causes it to have TF and SF values of 1.82 and 0.55 in power-on operation, as stated in Chapter 13.

When the output shaft of the transmission is stalled, the torque factors of the converter and gear set 1 are the same as when in the Drive position. This is true since the sun gear is stationary when the turbines are stationary whether the band is engaged or not. However, after the turbines start to rotate, the Held sun gear in the Low position causes the turbine 1 speed to be at all times (Condition 3, Chart 1),

$$S_{turb\,1} = 1.6\,S_{turb\,2}$$

The torque factors of the transmission, then, are 1.82 times those in the Drive position when the output shaft of the transmission is stalled. Likewise, the transmission torque factor is 1.82 times the torque factor of the converter and gear set 1 under nonstalled conditions until the slip of turbine 1 becomes very small. This slip becomes zero when the turbine 2 speed is 0.625 S_{imp}, and it is a negative slip of nearly 60 per cent when turbine 2 approaches zero slip. Thus, turbine 1 acts as an impeller and the torque and power for its impeller action are derived from turbine 2 when the speed of turbine 2 is greater than 0.625 S_{imp}; that is, when its slip is less than 37.5 per cent.

The transmission TF and SF values are:

$TF_{trans} = 6.2$ or 5.65 when the transmission is stalled, depending on the angle of the stator 2 vanes,

$TF_{trans} = 1.82$ to a value approaching one or the other of the above values in nonstalled operations, depending on the values of the slips of turbines 1 and 2,

$SF_{trans} = 0.55$ to 0.0, depending on the slip of turbine 2.

7. Low-Gear Power-Off Operation (In Low Position)

The conditions in the converter and gear set 1 are the same as in power-off operation when in the Drive position except that the sun gear is Held by the band. The speed of turbine 1, then, always is 1.6 times that of turbine 2 and 1.6 times that of the "new" input shaft of gear set 1. The converter acts as a fluid clutch with a torque factor which is substantially 1.0.

The TF and SF of gear set 2 in power-off operation (Chapter 13) are 0.55 and 1.82.

Turbines 1 and 2 act as impellers and the impeller acts as a turbine (or runner). Since turbine 1 rotates at a speed which is 1.6 times the speed of the "new" input shaft of gear set 1, its effectiveness in transmitting torque to the "new" turbine and the engine is much greater than it would be if its speed were equal to that of the "new" input shaft. The effectiveness of turbine 1 in power-off operation when in the Drive position is zero or substantially so. Thus, greater torque is transmitted to the engine and the engine is driven at a greater speed as the result of the use of the band to Hold the sun gear. Therefore, the "braking power" is greater. This increase is augmented by the torque and speed factors of gear set 2. The torque and speed factors of the transmission are

$$TF_{trans} = 0.55, \text{ and}$$
$$SF_{trans} = 1.82 \text{ to } 0.0$$

8. Reverse-Gear Power-On Operation

The conditions of the converter and gear set 1 are the same as those when in the Drive position. Therefore, the TF and SF of these components of the transmission are the same in the two positions.

Gear set 2 has the TF and SF values stated in Chapter 13, which are −1.82 and −0.55.

The TF and SF of the transmission, then, are the same as in Low-gear power-on operations except that the signs of the TF and SF are negative when in Reverse gear.

9. Reverse-Gear Power-Off Operation

The conditions in the converter and gear set 1 are the same as when in the Drive position. The SF and TF of gear set 2 are −1.82 and −0.55. Then,

$$TF_{trans} = -0.55, \text{ and}$$
$$SF_{trans} = -1.82 \text{ to } 0.0$$

10. Neutral-Gear Operation (In Neutral or Park Position)

The conditions of the converter and gear set 1 are the same as when in the Drive position. The SF and TF of gear set 2 are 0.0, as stated in Chapter 13. Therefore, the SF and TF of the transmission are 0.0.

11. Control System Considerations

The engaging and disengaging of the band in Fig. 30 may or may not change the ratio of the speeds of turbines 1 and 2. They have little or no effect on the natures of the control problems involved in the shifting of the gears of gear set 2. The natures of these problems, then, are substantially the same as those in § 8 of Chapter 13 insofar as shifts between Low and High gear are concerned. The other shifts of that transmission do not occur in the transmission in Fig. 30. The control problems of gear set 2 in Fig. 30, therefore, are not included in the following.

When in High gear, the ratio of the speeds of turbines 1 and 2 may have any value from substantially 1.0 to 1.6. In power-on operation it is 1.6 and the sun gear is Held by the overrunning clutch when the automobile is being started, and it may have this value at higher automobile speeds when the acceleration and/or road conditions demand a large torque from gear set 1. If a power-on change from High to Low gear is made under these conditions (when the speed ratio is 1.6), no control problems are involved with respect to the band shown in Fig. 30 since the only change is in the means used in Holding the sun gear. If a power-on change from High to Low gear is made when the turbine speed ratio is less than 1.6, the engagement of the band changes the ratio to 1.6. The control system should limit the rate of engagement to one which does not subject turbine 1, the gear set and the band to the excessive torques that may be required for a very rapid acceleration of the turbine. In this connection, the shift in gear set 2 increases the speed of turbine 2 by a factor of 1.82, so the increase in the speed of turbine 1 is affected by this factor in addition to the factor (1.6 or less) which results from the engagement of the band in Fig. 30. A change from Low to High gear in power-on operation changes the means of Holding the sun gear, or releases the sun gear so that it can rotate freely in the forward direction. No problems are presented to the control system by either of these changes.

In power-off operation in High gear, turbines 1 and 2 rotate at substantially the same speed. When a shift to Low gear is made, the Holding action of the band increases the ratio of the speeds of turbines 1 and 2 to 1.6, and the speed of turbine 1 increases by a factor of 2.91. As in the power-on shift, the engagement of the band should be controlled so that the acceleration of turbine 1 is not too great. No control problems are involved in power-off shifts from Low to High gear since the status of the sun gear is changed from Held to Free and turbine 1 is free to change speed at rates which are determined by the fluid and inertia forces acting on it.

18

Cruise-O-Matic and Merc-O-Matic Transmissions

1. General Description

The control levers of these transmissions have six positions: Drive 1, Drive 2, Low, Reverse, Neutral, and Park. In the *D1* position, the transmissions shift from 1st to 2nd to 3rd gear, and vice versa, or from 3rd to 1st gear. In the *D2* position the shifting is from 2nd to 3rd gear, and vice versa. When in the other positions, the transmissions do not shift. The terms 1st gear, Low gear, etc., and the conditions in the Park position are discussed in Chapter 2, § 4. The throttle control and a control operated by the manifold vacuum serve as transmission control devices, as described in Chapter 10.

Figure 31 illustrates a transmission which is the same as or similar to Cruise-O-Matic and Merc-O-Matic transmissions. The gear set is similar to that shown in Fig. 17, but sun gear 1 is larger than sun gear 2. These transmissions differ from those described in other chapters which employ this gear set in that they provide three forward speeds instead of two.

A three-element torque converter with a maximum *TF* of 2.0 is employed in the power train between the engine and the gear set. Overrunning clutches, described in Chapter 2, § 6, are used.

Section 4 of Chapter 10 provides information pertaining to the *TF* and *SF* values as they are stated in this chapter.

2. Numbers of Gear Teeth

The numbers of teeth on the gears are assumed to be:

Sun gear 1	—	36
Planet pinions 1	—	18
Ring gear	—	72
Sun gear 2	—	30
Planet pinions 2	—	15

169

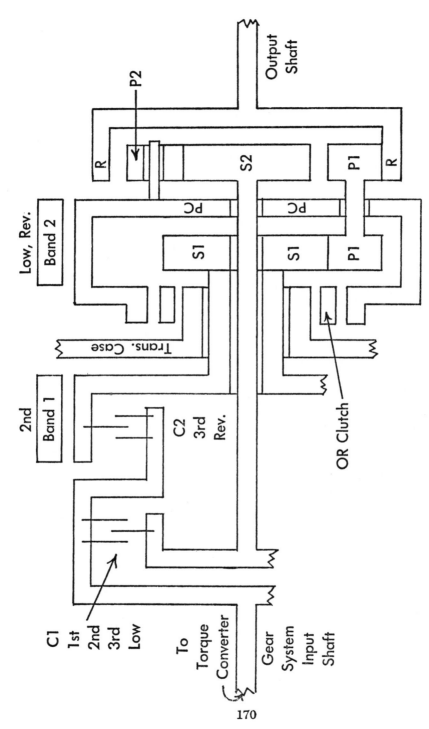

FIG. 31. Cruise-O-Matic and Merc-O-Matic transmissions. (1) 1st, Low, etc., indicate band or clutch is engaged in 1st gear, etc. (2) OR clutch Holds when planet carrier tends to rotate in reverse direction. (3) Gear set is similar to one shown in Fig. 17. (4) Bands are connected to transmission case.

170

3. 1st-Gear Power-On Operation (In Drive 1 Position)

Clutch 1 is engaged.

Torque is transmitted by sun gear 2 to pinions 2 and 1. The torque on the planet carrier is in the reverse direction, so the overrunning clutch Holds the carrier. Sun gear 1 is Free since clutch 2 and band 1 are disengaged. The gear set is in Condition 24 in Chart 4 and

$$TF = T_r/T_{s2} = 2.40, \text{ and}$$
$$SF = 0.4167$$

Since the maximum TF of the converter is 2.0,

$$TF_{\text{trans}} = 4.80 \text{ to } 2.40, \text{ and}$$
$$SF_{\text{trans}} = 0.4167 \text{ to } 0.0$$

4. 1st-Gear Power-Off Operation (In Drive 1 Position)

Clutch 1 is engaged.

If the "new" input shaft tends to drive the "new" output shaft, the load presented by the engine tends to limit the speed of that output shaft, or in effect tends to Hold the sun gear 2 shaft. When sun gear 2 is Held (Condition 16, Chart 4), the output torque of the planet carrier shaft is (Tq_{in}) $(1 - T_{s2}/T_r)$ which is in the forward direction, so the overrunning clutch overruns. Since sun gear 2 is partially Held by the engine load, the clutch overruns until the driving torque of the engine provides a reverse-direction torque on the planet carrier; that is, until a power-on condition exists. In power-off operation, then, the planet carrier and sun gear 1 are Free, the gear set is in one of Conditions 25–36 in Chart 4, and the TF and SF of the transmission are 0.0.

The transmission shifts to 2nd gear at about 8 mph when the throttle is closed.

5. Low-Gear Power-On Operation (In Low Position)

Clutch 1 and band 2 are engaged.

Band 2 Holds the planet carrier which is Held by the overrunning clutch in 1st-gear power-on operation. Therefore, the TF and SF of the transmission are the same as when in 1st gear.

6. Low-Gear Power-Off Operation (In Low Position)

Clutch 1 and band 2 are engaged.

Since band 2 Holds the planet carrier and sun gear 1 is Free, the condition of operation is the inverse of the condition in § 3, and the TF and SF of the gear set are

$$TF = T_{s2}/T_r = 0.4167, \text{ and}$$
$$SF = 2.40$$

171

Then,

$$TF_{\text{trans}} = 0.4167, \text{ and}$$
$$SF_{\text{trans}} = 2.40 \text{ to } 0.0$$

7. 2nd-Gear Power-On Operation (In Drive 1 or 2 Position)

Clutch 1 and band 1 are engaged.

Torque is transmitted from sun gear 2 to pinions 2 and 1 and to sun gear 1 which is Held. Pinions 1 rotate around sun gear 1 in the forward direction. The torque on the planet carrier is in the forward direction, so the overrunning clutch overruns and the planet carrier is Free. The gear set is in Condition 12 in Chart 4 and

$$TF = (1 + T_{s1}/T_{s2})/(1 + T_{s1}/T_r) = 1.47, \text{ and}$$
$$SF = 0.68$$

Then, in the Drive 1 position,

$$TF_{\text{trans}} = 1.47 \text{ to a higher value, and}$$
$$SF_{\text{trans}} = 0.68 \text{ to a lower value}$$

In the Drive 2 position,

$$TF_{\text{trans}} = 2.94 \text{ to } 1.47, \text{ and}$$
$$SF_{\text{trans}} = 0.68 \text{ to } 0.0$$

8. 2nd-Gear Power-Off Operation (In Drive 1 or 2 Position)

Clutch 1 and band 1 are engaged.

The torque from the "new" input shaft flows through pinions 1 to sun gear 1 which is Held by band 1. Therefore, pinions 1 rotate around sun gear 1 in the forward direction and the overrunning clutch overruns. The planet carrier is Free, the gear set is in Condition 11 in Chart 4 which is the inverse of the condition in power-on operation, and

$$TF = 0.68, \text{ and}$$
$$SF = 1.47$$

Then,

$TF_{\text{trans}} = 0.68,$
$SF_{\text{trans}} = 1.47$ to a lower value when in the $D1$ position, and
$SF_{\text{trans}} = 1.47$ to 0.0 when in the $D2$ position.

When the automobile speed is above approximately 20 mph, the transmission shifts to 3rd gear if the throttle is closed.

9. 3rd-Gear Power-On Operation (In Drive 1 or 2 Position)

Clutches 1 and 2 are engaged.

The clutches lock the sun gear shafts to one another. The planet carrier is Free since it rotates in the forward direction. One of Conditions 49–60 in Chart 4 exists and the TF and SF are 1.0, so

$$TF_{trans} = 1.0 \text{ to a higher value, and}$$
$$SF_{trans} = 1.0 \text{ to a lower value}$$

10. 3rd-Gear Power-Off Operation (In Drive 1 or 2 Position)

Clutches 1 and 2 are engaged.

The gear set is in one of Conditions 49–60 in Chart 4 and its TF and SF are 1.0, so

$$TF_{trans} = 1.0, \text{ and}$$
$$SF_{trans} = 1.0 \text{ to a lower value}$$

11. Reverse-Gear Power-On Operation

Clutch 2 and band 2 are engaged.

Sun gear 2 is Free and the planet carrier is Held. The gear set is in Condition 1 in Chart 4, and

$$TF = -T_r/T_{s1} = -2.0, \text{ and}$$
$$SF = -0.50$$

Then,

$$TF_{trans} = -4.0 \text{ to } -2.0, \text{ and}$$
$$SF_{trans} = -0.50 \text{ to } 0.0$$

12. Reverse-Gear Power-Off Operation

Clutch 2 and band 2 are engaged.

The direction of rotation of the "new" input shaft is the reverse direction when the direction of rotation of the engine crankshaft is considered to be the forward direction. The condition of operation is the inverse of that in power-on operation, so

$$TF_{gear\ set} = -0.50,$$
$$SF_{gear\ set} = -2.0,$$
$$TF_{trans} = -0.50, \text{ and}$$
$$SF_{trans} = -2.0 \text{ to } 0.0$$

Since the TF and SF are negative and the "new" input shaft rotates in the reverse direction with respect to the direction of the crankshaft, the "new" output shaft rotates in the forward direction.

13. Neutral-Gear Operation (In Neutral or Park Position)

No clutch or band is engaged.

Sun gears 1 and 2 are Free, and the planet carrier is Free insofar as forward-direction rotation is concerned. The gear set is in one of Conditions 25–36 in Chart 4, and the TF and SF of the transmission are 0.0.

173

14. Control System Considerations

The pattern of clutch and band engagements in the drive transmission control positions and gears, and the corresponding gear system torque factors, are shown in Table 13. The symbol, E, indicates engaged, and H or OR means that the overrunning clutch Holds or overruns. NI means that the overrunning clutch is "not involved" since band 2 Holds the planet carrier, although it may assist the band in Holding the carrier.

TABLE 13. CONTROL SYSTEM PATTERN

CP	Gear	Clutch 1	Clutch 2	Band 1	Band 2	OR Clutch P-On	OR Clutch P-Off	Gear Syst TF P-On	Gear Syst TF P-Off
D1	1st	E				H	OR	2.40	0.000
D1,D2	2nd	E		E		OR	OR	1.47	0.680
D1,D2	3rd	E	E			OR	OR	1.00	1.000
L	Low	E			E	NI	NI	2.40	0.4167
R	Rev.		E		E	NI	NI	−2.00	−0.5000

The control system should prevent simultaneous engagement of clutch 2 and band 1 to significant extents during shifts between 2nd and 3rd gears since the torques of these would "fight" one another. The same pertains to clutch 2 and band 2 in shifts from 3rd to Low gear since simultaneous engagement of these and clutch 1 tends to place the transmission in 3rd gear and Reverse gear simultaneously. Likewise, simultaneous engagements of bands 1 and 2 during shifts from 2nd to Low gear tend to place the transmission in 2nd gear and in Low gear.

As in some of the preceding chapters, in power-on up-shifts from 2nd to 3rd gear the period of simultaneous disengagement of band 1 and clutch 2 should be no more than a very short one in order to prevent an increase in the speed of the engine, and completion of the engagement of clutch 2 should be at a rate which prevents a surge in the speed of the automobile as the result of excess engine speed.

Also, as in preceding Chapters, a period of simultaneous disengagement of the elements involved in power-on down-shifts from 3rd to 2nd gear, or from either of these to Low gear, followed by rapid engagement of band 1 or 2, would be desirable in order that the engine speed may increase prior to completion of the shift to the new gear. Power-off shifts from 3rd, 2nd or Low gear to 1st gear include no engine speed problems since the power-off TF in 1st gear is zero.

19

Studebaker Automatic Transmissions

1. General Description

The control lever has five positions: Drive, Low, Reverse, Neutral, and Park. The throttle control serves as a transmission control, as described in Chapter 10.

When in the *D* position, the transmission shifts from 2nd to 3rd gear, and vice versa. When in the other positions, the transmission does not shift. The terms 1st gear, Low gear, etc., and the conditions in the Park position are discussed in Chapter 2, § 4.

Figure 32 illustrates a transmission which is the same as or similar to a Studebaker transmission. The three-element converter has a maximum *TF* of 2.1. Three overrunning clutches, described in Chapter 2, § 6, are used.

The transmission is of unusual design in that the converter and gears are not used in the power train when the transmission is in 3rd gear. Another unusual item is the "Hill-Holding" feature which prevents the automobile from rolling backward when it is stopped on an up-hill grade with the transmission control in the Drive position and the engine running.

Section 4 of Chapter 10 provides information pertaining to the *SF* and *TF* values as they are stated in this Chapter.

2. Numbers of Gear Teeth

The numbers of teeth on the gears are assumed to be:

	Set 1	Set 2
Sun gear	56	40
Planet pinions	18	26
Ring gear	92	92

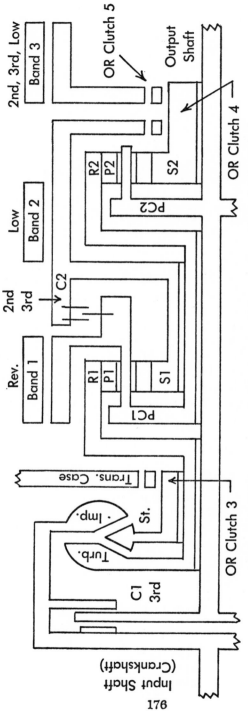

FIG. 32. Studebaker automatic transmission. (1) 2nd, Low, etc., indicate band or clutch is engaged in 2nd gear, etc. (2) OR clutch 3 or 5 Holds when stator or sun gear 2 tends to rotate in reverse direction. OR clutch 4 Holds when its outer race tends to rotate in reverse direction with respect to sun gear 2, or when sun gear 2 tends to rotate in forward direction with respect to the outer race. (3) Bands are connected to transmission case.

176

3. 2nd-Gear Power-On Operation (In Drive Position)

This is the starting gear when the control is in the Drive position. Clutch 2 and band 3 are engaged.

The power flow is from the input shaft (crankshaft) to the torque converter impeller and turbine and ring gear 1. Clutch 2 locks planet carrier 1 to sun gear 1 so the torque from the converter passes through gear set 1 with a TF of 1.0. This torque is placed on ring gear 2, which causes a forward torque on planet carrier 2 and a reverse torque on sun gear 2. Overrunning clutch 5 Holds the sun gear stationary and the outer race of overrunning clutch 4 overruns the shaft of sun gear 2. Gear set 2 is in Condition 4 of Chart 1 and

$$TF = 1 + T_s/T_r = 1.435, \text{ and}$$
$$SF = 0.697$$

Then,

$$TF_{\text{trans}} = 3.01 \text{ to } 1.435, \text{ and}$$
$$SF_{\text{trans}} = 0.697 \text{ to } 0.0$$

4. 2nd-Gear Power-Off Operation (In Drive Position)

Clutch 2 and band 3 are engaged.

The "new" input shaft drives planet carrier 2 in the direction of rotation of the crankshaft. Gear set 2 is in Condition 6 in Chart 3, and

$$TF_1 = Tq_{\text{out } s}/Tq_{\text{in}} = 1/(1 + T_r/T_s) = 0.303, \text{ and}$$
$$TF_2 = Tq_{\text{out } r}/Tq_{\text{in}} = 1/(1 + T_s/T_r) = 0.697$$

Overrunning clutch 4 Holds and overrunning clutch 5 overruns, so a torque of $0.303\, Tq_{\text{in}}$ is applied to the combination of sun gear 1 and planet carrier 1. The torque on ring gear 2, $0.697\, Tq_{\text{in}}$, is applied to planet carrier 1. Since planet carrier 1 and sun gear 1 are locked together, the TF of gear set 1 is 1.0 and the sum of the above torques, or Tq_{in}, is applied to ring gear 1 and to the converter turbine. The TF of the converter in power-off operation is 1.0, so

$$TF_{\text{trans}} = 1.0, \text{ and}$$
$$SF_{\text{trans}} = 1.0 \text{ to } 0.0$$

The transmission shifts to 3rd gear when the automobile speed is above about 18 mph and the throttle is closed.

5. 2nd-Gear Hill-Holding Operation (In Drive Position)

Clutch 2 and band 3 are engaged.

The torque of the "new" input shaft is in the direction opposite to the direction of rotation of the crankshaft since the automobile tends to roll backward down the hill. The torque factors of gear set 2 are the same as in § 4. In this instance, the torque on sun gear 2 causes this sun gear to be Held by overrunning clutch 5. The torque on ring gear 2 is transmitted through clutch 2 to the outer race of overrunning clutch 4. Since this torque is in the reverse direction (with respect to the crankshaft rotation), overrunning clutch 4 Holds and tends to turn sun gear 2, but this turning is prevented by clutch 5. Thus, the "new" input shaft cannot rotate in the direction opposite to that of the crankshaft, and the automobile is held stationary

6. 3rd-Gear Power-On Operation (In Drive Position)

Clutches 1 and 2 and band 3 are engaged.

Clutch 1 resembles a manually operated clutch but it is operated automatically by hydraulic pressure. The engagement of this clutch connects the engine crankshaft to the output shaft of the transmission so the TF and SF of the transmission are 1.0 and the converter has no effect on these values.

The friction within the converter and gear sets tends to cause the turbine to rotate slower than the impeller and crankshaft. Thus, ring gear 2 tends to rotate slower than the crankshaft, and as a result sun gear 2 tends to rotate faster than the crankshaft. The resultant of these relative speeds of the ring and sun gears would be a component of rotation of the outer race of overrunning clutch 4 in the reverse direction with respect to the rotation of sun gear 2. This component cannot exist since clutch 4 Holds when the outer race tends to rotate in the reverse direction with respect to the rotation of the sun gear. Therefore, the outer race, gear sets 1 and 2, and the turbine rotate at the speed of the impeller and crankshaft. This reduces the power loss in the converter and gear system to a value below that which would occur if the gear system were not locked so as to rotate as a unit and if the turbine were permitted to rotate slower than the impeller.

7. 3rd-Gear Power-Off Operation (In Drive Position)

Clutches 1 and 2 and band 3 are engaged.

The "new" input shaft drives the crankshaft through clutch 1 and the TF and SF of the transmission are 1.0.

The torque on sun gear 2 is transmitted through overrunning clutch 4 and clutch 2 so gear sets 1 and 2 and the turbine rotate at the speed of the "new" input shaft and crankshaft.

8. Low-Gear Power-On Operation (In Low Position)

Bands 2 and 3 are engaged.

Band 2 Holds sun gear 1 and gear set 1 is in Condition 4 in Chart 1, so

$$TF_{\text{gear set 1}} = 1 + T_s/T_r = 1.609$$

178

The output torque of gear set 1 is the input torque of ring gear 2. This torque on ring gear 2 tends to rotate sun gear 2 in the reverse direction, but clutch 5 Holds the sun gear. Gear set 2 is in Condition 4 in Chart 1 and

$$TF_{\text{gear set 2}} = 1 + T_s/T_r = 1.435$$

Then,

$$TF_{\text{gear syst}} = (1.609)\,(1.435) = 2.31,$$
$$SF_{\text{gear syst}} = 0.433,$$
$$TF_{\text{trans}} = 4.85 \text{ to } 2.31, \text{ and}$$
$$SF_{\text{trans}} = 0.433 \text{ to } 0.0$$

9. Low-Gear Power-Off Operation (In Low Position)

Bands 2 and 3 are engaged.

The "new" input shaft rotates planet carrier 2 in the forward direction. This tends to rotate sun gear 2 in the forward direction, but band 2 and clutch 4 Hold the sun gear. Gear set 2 is in Condition 3 in Chart 1, and

$$TF_{\text{gear set 2}} = 1/(1 + T_s/T_r) = 1/1.435$$

The output torque of gear set 2 is applied to planet carrier 1, gear set 1 is in Condition 3 in Chart 1 since sun gear 1 is Held by band 2, and

$$TF_{\text{gear set 1}} = 1/(1 + T_s/T_r) = 1/1.609$$

Then,

$$TF_{\text{gear syst}} = (1/1.435)\,(1/1.609) = 0.433,$$
$$TF_{\text{trans}} = 0.433, \text{ and}$$
$$SF_{\text{trans}} = 2.31 \text{ to } 0.0$$

10. Reverse-Gear Power-On Operation

Band 1 is engaged.

Torque from the converter is applied to ring gear 1 and gear set 1 is in Condition 2 in Chart 1, so

$$TF_{\text{gear set 1}} = -T_s/T_r = -0.609$$

The output torque of gear set 1 is applied through clutch 4 to sun gear 2 since the outer race torque is in the reverse direction. Gear set 2 is in Condition 5 in Chart 1, and

$$TF_{\text{gear set 2}} = 1 + T_r/T_s = 3.3$$

Then,

$$TF_{\text{gear syst}} = (3.3)\,(-0.609) = -2.01,$$
$$TF_{\text{trans}} = -4.22 \text{ to } -2.01, \text{ and}$$
$$SF_{\text{trans}} = -0.50 \text{ to } 0.0$$

11. Reverse-Gear Power-Off Operation

Band 1 is engaged.

The "new" input shaft rotates in the reverse direction with respect to the direction of rotation of the crankshaft. Its rotation drives sun gear 2 in the reverse direction against substantially no torque since neither overrunning clutch prevents this rotation, band 3 being disengaged. Therefore, substantially no torque is transmitted to ring gear 2 and the remainder of the transmission and the TF and SF of the transmission are 0.0.

12. Neutral-Gear Power-On Operation (In Neutral or Park Position)

No clutch or band is engaged.

Engine torque flows to ring gear 1. This torque tends to drive planet carrier 1 and ring gear 2 in the forward direction. It also tends to drive sun gears 1 and 2 in the reverse direction since clutch 4 Holds when its outer race rotates in the reverse direction with respect to sun gear 2. Gear set 1 is in Condition 2 in Chart 3, and

$$TF_1 = Tq_{\text{out s}}/Tq_{\text{in}} = -T_s/T_r = -0.609,$$
$$TF_2 = Tq_{\text{out pc}}/Tq_{\text{in}} = 1 + T_s/T_r = 1.609, \text{ and}$$
$$Tq_{\text{out s}}/Tq_{\text{out pc}} = -0.609/1.609$$

The speeds of sun gear 1 and planet carrier 1, with any particular ring gear 1 torque, are determined by the torque-vs.-speed characteristics of the loads connected to the shafts of those elements (Chapter 5, § 4).

Gear set 2 may appear to be in Condition 5 in Chart 3 but it is not in that condition since the two input shafts rotate in opposite directions.

Ring gear 2 tends to rotate planet carrier 2 in the forward direction and sun gear 2 tends to rotate the carrier in the reverse direction. The load connected to the output shaft tends to prevent rotation of the carrier. The carrier speed is zero when

$$(S_r) (1/[1 + T_s/T_r]) + (-S_s) (1/[1 + T_r/T_s]) = 0.0$$

where all of the quantities pertain to gear set 2. The carrier speed is zero, then, when $S_r/S_s = T_s/T_r = 0.435$. When this speed ratio exists, the torques on the ring and sun gears of gear set 2 are used to rotate pinions 2 on their axes and the torque applied to the output shaft is zero. The output torques of the sun gear and planet carrier of gear set 1, then, are those required by friction within the gear system and the ratio of these is stabilized at the value above. If the torque provided by ring gear 2 and planet carrier 1 should increase to a value above its stabilized value, the speed of ring gear 2 would increase. Assuming that planet carrier 2 remains stationary, the speed of sun gear 2 will increase and the torque

180

requirement placed on sun gear 1 will decrease since the torque for the sun gear 2 speed is derived from the ring gear. The ratio of $Tq_{out\ s}/Tq_{out\ pc}$ of gear set 1 will be less than the value it must have (Chapter 5, § 4), so the torque provided by ring gear 2 and planet carrier 1 decreases and that provided by sun gears 2 and 1 increases and the torque ratio is restored to the value it must have. A similar adjustment of this ratio would occur if the torque provided by sun gears 2 and 1 should increase to a value above its stabilized value.

From the above,

$$TF_{trans} = (\text{substantially}) \ 0.0, \text{ and}$$
$$SF_{trans} = 0.0$$

13. Neutral-Gear Power-Off Operation (In Neutral Position)

No clutch or band is engaged.

If the "new" input shaft rotates in the direction opposite to that of the crankshaft, the conditions pertaining to sun gear 2 are the same as in § 11, so the TF and SF of the transmission are 0.0.

If the "new" input shaft rotates in the same direction as the crankshaft, planet carrier 2 tends to rotate sun gear 2 and ring gear 2 in the forward direction. Overrunning clutch 4 Holds since there is no torque opposing the rotation of sun gear 2 except that provided through the overrunning clutch. Then, planet carrier 2 tends to drive sun gear 1 and planet carrier 1 in the forward direction. Forward rotation of planet carrier 1 tends to rotate ring gear 1 in the forward direction and forward rotation of sun gear 1 tends to rotate the ring gear in the reverse direction. A situation similar to that in Neutral-gear power-on operation exists. As in that case, the speeds of the sun gear and planet carrier of gear set 1 automatically are adjusted so that the torque applied to ring gear 1 and the turbine is zero or substantially so. Therefore, the TF and SF of the transmission are 0.0.

14. Control System Considerations

The pattern of clutch and band engagements in the drive transmission control positions and gears, and the corresponding gear system torque factors, are shown in Table 14. The symbol, E, indicates engaged, and H or OR means that the overrunning clutch Holds or overruns. NI means that the overrunning clutch is "not involved" in the transmission of torque through the transmission. In Low gear power-on operation clutch 4 is not involved since sun gear 2 tends to rotate in the reverse direction, which would cause clutch 4 to overrun, but clutch 5 prevents this rotation of the sun gear. Overrunning clutch 5 is not involved in Low-gear power-off operation since sun gear 2 tends to rotate in the forward direction, which would cause clutch 5 to overrun, but clutch 4

181

prevents this rotation of the sun gear. Clutch 5 is not involved in Reverse-gear operation in the sense that it does not transmit torque to the transmission case when band 3 is disengaged.

TABLE 14. CONTROL SYSTEM PATTERN

CP	Gear	Clutch 1	Clutch 2	Band 1	Band 2	Band 3	OR Clutch 4 P-On	OR Clutch 4 P-Off	OR Clutch 5 P-On	OR Clutch 5 P-Off	Gear Syst TF P-On	Gear Syst TF P-Off
D	2nd		E			E	OR	H	H	OR	1.435	1.000
	3rd	E	E			E	H	H	OR	OR	1.000	1.000
L	Low				E	E	NI	H	H	NI	2.310	0.433
R	Rev.			E			H	OR	NI	NI	−2.010	0.000

A power-on shift from 2nd to 3rd gear with no change in the acceleration of the automobile during the shift involves a change in the engine speed by a factor of approximately 1/1.435. Therefore, the completion of the engagement of clutch 1 should be at a rate which permits the engine speed to change this amount during the engagement. In this connection, the torque converter is not in the power train in 3rd gear, so it does not assist clutch 1 in the speed adjustment. Power-on shifts from 3rd to 2nd gear involve no control problems other than a reasonably slow disengagement of clutch 1 since the engine otherwise is free to change to its new speed as clutch 1 disengages.

Power-off shifts from 2nd to 3rd gear, and vice versa, involve only small adjustments in the speed of the engine since the SF of the gear system is 1.0 in each of the gears. Thus, the only adjustments in these cases are those needed because of the slips of the converter.

Clutch 2 and band 2 should not be engaged simultaneously to significant extents in shifts between 2nd or 3rd gear and Low gear since the torques of these elements would "fight" one another. Likewise, clutch 1 and band 2 should not be engaged simultaneously since this would tend to place the transmission in 3rd and Low gears at the same time. As in other cases in which none of the torque factors involved are zero, power-off down-shifts should permit no more than very short periods of simultaneous disengagement of the clutches and bands which are changed in order that the engine speed will not have time to decrease appreciably during the shift. Also as in other transmissions, in power-on down-shifts a period of simultaneous disengagement followed by rapid engagement would be desirable. In power-on up-shifts from Low to 2nd gear, the control system should time the disengagement of band 2 and engagement of clutch 2 so as to prevent an appreciable increase in engine speed, and complete the engagement of clutch 2 at a rate which prevents a surge in the speed of the automobile as the result of the increase in speed factor of the transmission.

20

Hydra-Matic Transmissions

1. General Description

Hydra-Matic transmissions have been constructed in different ways and used in several makes of automobiles. Figure 33 illustrates a transmission which is the same as or similar to one model of the Hydra-Matic transmissions. When in the D position, the transmission shifts from 1st to 2nd, 3rd, and 4th gears and down-shifts in the ways indicated in Chapter 10. In the S position, the shifting is limited to 1st, 2nd, and Third gears. In the L position, it is from 1st to Second gear, and vice versa. The transmission does not shift when in the R, N, or P position. The terms 1st gear, Second gear, etc., and the conditions in the Park position are discussed in Chapter 2, § 4. The throttle control serves as a transmission control device, as described in Chapter 10.

The transmission shown in Fig. 33 is of unusual design in that: (1) four forward speeds are provided, (2) two fluid clutches are used, (3) the principal fluid clutch (clutch 6) is not coupled directly to the engine crankshaft, (4) the second fluid clutch has no fluid in it when the transmission is operating in 1st and 3rd gears, (5) a triple planetary gear system is used, (6) the torque flow is divided between two types of paths, one mechanical and the other mechanical and hydraulic, when operating in 2nd, 3rd, and 4th gears, and in 4th gear it is divided in two sets of these types of paths.

The second fluid clutch (clutch 5) is used to (substantially) lock together two elements of a planetary gear set in lieu of a mechanical multi-disc clutch which usually is employed for this purpose.

Two overrunning clutches, described in Chapter 2, § 6, are used.

Section 4 of Chapter 10 provides information pertaining to the SF and TF values as they are stated in this Chapter.

2. Numbers of Gear Teeth

The numbers of teeth on the gears are assumed to be:

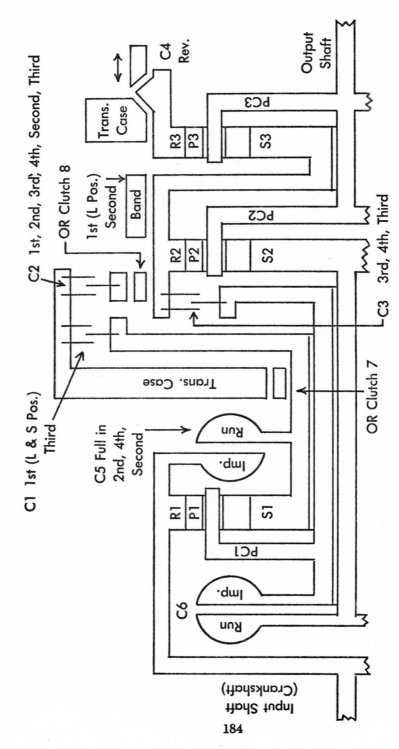

FIG. 33. Hyrdra-Matic transmission. (1) 1st, 2nd, etc., indicate band or clutch engaged in 1st gear, etc. (2) OR clutch 7 or 8 Holds when sun gear 1 or ring gear 2 tends to rotate in reverse direction. (3) Band is connected to transmission case.

184

	Set 1	Set 2	Set 3
Sun gear	30	41	28
Planet pinions	18	13	20
Ring gear	67	67	68

3. 1st-Gear Power-On Operation (In Drive Position)

Clutch 2 is engaged. Fluid clutch 5 has no fluid in it and therefore is "disengaged."

Torque on ring gear 1 from the engine tends to rotate sun gear 1 in the reverse direction, so overrunning clutch 7 Holds the sun gear. Gear set 1 is in Condition 4 in Chart 1, and

$$TF_{\text{gear set 1}} = 1 + T_s/T_r = 1.45$$

The TF of the principal fluid clutch (clutch 6) is substantially 1.0. Then all of the output torque of gear set 1 is applied to the sun gear of gear set 2, which is in Condition 5 in Chart 1 since ring gear 2 is Held by overrunning clutch 8 and clutch 2. Then,

$$TF_{\text{gear set 2}} = 1 + T_r/T_s = 2.63, \text{ and}$$
$$TF_{\text{trans}} = (1.45)(2.63) = 3.81$$

With a small amount of slip in clutch 6,

$$SF_{\text{trans}} = 0.262$$

which corresponds with a speed reduction ratio of 3.82.

Then, with the range of slips that occur in clutch 6,

$$SF_{\text{trans}} = 0.262 \text{ to } 0.0$$

4. 1st-Gear Power-Off Operation (In Drive Position)

The clutch and band conditions are the same as in § 3.

The "new" input shaft tends to drive the sun and ring gears of gear set 2 in the forward direction with respect to the direction of rotation of the crankshaft. Overrunning clutch 8 overruns, so (substantially) no torque is placed on sun gear 2 and the remainder of the transmission. Therefore, the TF and SF of the transmission are 0.0.

The transmission shifts to 2nd gear when the automobile speed is above approximately 6 mph and the throttle is closed.

5. 2nd-Gear Power-On Operation (In Drive or S Position)

Clutch 2 is engaged. Fluid clutch 5 is full of fluid soon after the shift to 2nd gear is started, and therefore is "engaged."

Fluid clutch 5 (substantially) locks together the sun and ring gears of gear set 1 when the slip of this clutch is small. Gear set 1, then, (substantially) is in one of Conditions 13–18 in Chart 1, and

$$TF_{\text{gear set 1}} = 1.0$$

185

Since the TF of fluid clutch 6 is 1.0, all of the engine torque is applied to sun gear 2. Gear set 2 is in Condition 5 in Chart 1, and

$$TF_{\text{gear set 2}} = 1 + T_r/T_s = 2.63$$

Then,

$$TF_{\text{trans}} = 2.63, \text{ and}$$
$$SF_{\text{trans}} = 0.38$$

to a slightly lower value when the slips of fluid clutches 5 and 6 are small.

The change from 1st to 2nd gear results from the filling of fluid clutch 5. In 1st gear (§ 3), gear set 1 is considered as being in Condition 4 in Chart 1, and in 2nd gear it is considered in the above as being substantially in one of Conditions 13–18 in that chart when the slip of fluid clutch 5 is small. Gear set 1 also may be considered to be in Condition 5 in Chart 3 in 1st and 2nd gears and throughout the change from one gear to the other. When clutch 5 is empty, the input torque of the sun gear shaft is provided by overrunning clutch 7 and the transmission case. As the clutch fluid enters, the input torque is provided by the overrunning clutch and the fluid clutch runner, with the latter component becoming greater and greater until the overrunning clutch starts to overrun. After this time, all of the input torque of the sun gear is provided by the runner. In Condition 5 in Chart 3,

$$SF_2 = S_{\text{out}}/S_{\text{in r}} = (T_s/A + T_r)/(T_s + T_r)$$

where

$$A = S_{\text{in r}}/S_{\text{in s}},$$
$$TF_2 = Tq_{\text{out}}/Tq_{\text{in r}} = 1 + T_s/T_r, \text{ and}$$
$$TF_1 = Tq_{\text{out}}/Tq_{\text{in s}} = 1 + T_r/T_s$$

When in 1st gear, A is infinite if $S_{\text{in r}}$ is not zero since $S_{\text{in s}}$ is zero, so

$$SF_2 = SF_{\text{gear set 1}} = T_r/(T_s + T_r) = 0.69, \text{ and}$$
$$TF_2 = Tq_{\text{out}}/Tq_{\text{in r}} = TF_{\text{gear set 1}} = 1 + T_s/T_r = 1.45$$

which are the SF and TF values in § 3.

When overrunning clutch 7 overruns and the slip of clutch 5 approaches zero, A is slightly greater than 1.0. SF_2, then, is slightly less than 1.0. As the slip varies from 100 per cent to a small value, the SF of the gear set varies from 0.69 to slightly less than 1.0.

The ratio

$$Tq_{\text{in r}}/Tq_{\text{in s}} = T_r/T_s = 2.22$$

is a constant value (Chapter 5, § 2). The TF of clutch 5 is substantially 1.0 when the clutch drives sun gear 1, so

$$Tq_{\text{in r}}/Tq_{\text{in}} \text{ clutch 5} = 2.22$$

From Chapter 5, § 2,

$$Tq_{in\ r} + Tq_{in\ s} + Tq_{in\ pc} = 0.0, \text{ or}$$
$$Tq_{in} \text{ from engine} = Tq_{in\ r} + Tq_{in\ s} = Tq_{out\ pc}$$
$$= Tq_{in\ r} + Tq_{in\ r}/2.22 = Tq_{out\ pc}, \text{ or}$$
$$Tq_{in\ r} = (Tq_{in} \text{ from engine})/1.45 = 0.69\ (Tq_{in} \text{ from engine})$$

Thus, 69 per cent of the engine torque flows to the planet carrier through the ring gear, or through a mechanical path, and 31 per cent flows through a mechanical and hydraulic path, in going from the engine to the impeller of fluid clutch 6. From this impeller the torque flows through a mechanical and hydraulic path.

Since the TF of clutch 5 is substantially 1.0 when this clutch drives sun gear 1,

$$Tq_{out\ pc}/(Tq_{in\ r} + Tq_{in\ s}) = Tq_{out\ pc}/(Tq_{in} \text{ from engine}) = 1.0$$

and the TF of the combination of gear set 1 and clutch 5 is substantially 1.0 when the clutch has enough fluid to cause overrunning clutch 7 to overrun.

From the above, considering the possible slips of clutches 5 and 6,

$TF_{trans} = 3.81$ ($= TF$ in 1st gear) to 2.63 when clutch 5 has insufficient fluid to cause clutch 7 to overrun,

$TF_{trans} = 2.63$ when the clutch overruns, and

$SF_{trans} = 0.38$ to a lower value.

The lower value is 0.262, or the 1st-gear SF, when fluid clutch 5 has insufficient fluid to cause clutch 7 to overrun and when the slip of clutch 6 is very small. This condition occurs under light-load conditions immediately after the control system moves the gear-shift valve to the 2nd-gear position and before an appreciable amount of fluid has entered clutch 5. When clutch 5 has sufficient fluid to cause clutch 7 to overrun,

$$SF_{trans} = 0.38$$

to a lower value, depending on the slips of both of the fluid clutches.

6. 2nd-Gear Power-Off Operation (In Drive or S Position)

The conditions in gear set 2 are the same as when in 1st-gear power-off operation, so the TF and SF of the transmission are 0.0.

The transmission shifts to 3rd gear or Third gear when the automobile speed is above approximately 15 mph if the throttle is closed.

7. 3rd-Gear Power-On Operation (In Drive Position)

Clutches 2 and 3 are engaged. Fluid clutch 5 is empty.

Gear set 1 is in the same condition as when in 1st gear, and

$$TF_{gear\ set\ 1} = 1 + T_s/T_r = 1.45$$

187

Clutch 3 and fluid clutch 6 substantially lock together the ring and sun gears of gear set 2 when the slip of the fluid clutch is small, so

$$TF_{\text{gear set 2}} = 1.0$$

Then,

$$TF_{\text{trans}} = 1.45, \text{ and}$$
$$SF_{\text{trans}} = 0.69$$

to a slightly lower value when the slip of clutch 6 is small.

The torque from the engine to ring gear 2 flows through a mechanical path while that to sun gear 2 flows through a mechanical and hydraulic path. Gear set 2, while almost in one of Conditions 13–18 in Chart 1 when the slip of clutch 6 is small, is in Condition 5 in Chart 3, and

$$Tq_{\text{in r}}/Tq_{\text{in s}} = T_r/T_s = 1.63$$

The total torque on gear set 2 is

$$Tq_{\text{in gear set}} = Tq_{\text{in r}} + Tq_{\text{in s}} = Tq_{\text{out pe}}$$
$$= 1.63\, Tq_{\text{in s}} + Tq_{\text{in s}} = 2.63\, Tq_{\text{in s}}$$

Then,

$$Tq_{\text{in s}}/Tq_{\text{in gear set}} = 1/2.63 = 0.38$$

and 38 per cent of the torque flowing to gear set 2 flows through a mechanical and hydraulic path and 62 per cent flows through a mechanical path.

When the slip of clutch 6 varies through its normal range of values,

$$TF_{\text{trans}} = 1.45, \text{ and}$$
$$SF_{\text{trans}} = 0.69 \text{ to a lower value, depending on the slip}$$

8. 3rd-Gear Power-Off Operation (In Drive Position)

Clutch and band conditions are the same as in § 7.

Torque from the "new" input shaft tends to rotate the ring and sun gears of gear set 2 in the forward direction with respect to the direction of rotation of the crankshaft. Overrunning clutch 8, then, overruns. Forward-direction torque from ring gear 2 is placed on the shaft to planet carrier 1 and the impeller of fluid clutch 6. Forward-direction torque also is placed on the runner of this clutch by sun gear 2, which tends to produce a forward torque on the impeller. The forward torque on planet carrier 1 due to both of the above torques rotates sun gear 1 freely in the forward direction. Therefore, no torque is placed on ring gear 1 and the engine and the TF and SF of the transmission are 0.0.

The transmission shifts to 4th gear when the automobile speed is above approximately 22 mph if the throttle is closed.

9. 4th-Gear Power-On Operation (In Drive Position)

Clutches 2 and 3 are engaged and fluid clutch 5 is full of fluid.

When the slip of fluid clutch 5 is small, gear set 1 is substantially in one of Conditions 13–18 in Chart 1 and its TF and SF are approximately 1.0. Clutch 3 and fluid clutch 6 likewise place gear set 2 substantially in one of Conditions 13–18 in Chart 1 when the slip of this fluid clutch is small. Then, when the slips of both clutches are small,

$$TF_{trans} = 1.0, \text{ and}$$
$$SF_{trans} = \text{slightly less than } 1.0$$

When the slip of one or both of the clutches is not small, one of the speed factors of gear sets 1 and 2 (Condition 5, Chart 3) is

$$SF_2 = S_{out\ pc}/S_{in\ r} = (T_s/A + T_r)/(T_s + T_r)$$

where

$$A = \pm S_{in\ r}/S_{in\ s}$$

In each instance, the sun gear speed is less than the speed of the ring gear, so A is greater than 1.0. Then the speed factors of the gear sets are less than 1.0 by amounts that are determined by the slips of the clutches, and the speed factor of the transmission is reduced accordingly.

The torque factor of each of the combinations of gear set and fluid clutch is substantially 1.0, as in the case of gear set 1 and clutch 5 when in 2nd gear (§ 5). Therefore,

$$TF_{trans} = 1.0, \text{ and}$$
$$SF_{trans} = 1.0 \text{ to a lower value}$$

The division of the torque from the engine between a mechanical path and a mechanical and hydraulic path insofar as the paths to planet carrier 1 are concerned is the same as when in 2nd gear, or 69 and 31 per cent, respectively. Likewise, the division between planet carrier 1 and the output shaft is the same as when in 3rd gear, or 62 and 38 per cent through the mechanical path and the mechanical and hydraulic path, respectively.

10. 4th-Gear Power-Off Operation (In Drive Position)

The clutch and band conditions are the same as in § 9.

As in § 9, when the slips of the fluid clutches are small, gear sets 1 and 2 are substantially in one of Conditions 13–18 in Chart 1, and

$$TF_{trans} = 1.0, \text{ and}$$
$$SF_{trans} = \text{slightly less than } 1.0$$

189

If the slips of the clutches are not small, the gear sets are considered to be in Condition 6 in Chart 3, and

$$SF_1 = S_{\text{out } s}/S_{\text{in pc}} = (T_s + T_r)/(T_s + B\,T_r), \text{ and}$$
$$SF_2 = S_{\text{out } r}/S_{\text{in pc}} = (T_s + T_r)/(T_s/B + T_r)$$

where

$$B = S_{\text{out } r}/S_{\text{out } s}$$

The torque of the "new" input shaft on planet carrier 2 divides between the ring and sun gears of gear set 2 with a ratio of T_r/T_s. Then, the torque on the shaft from ring gear 2 to planet carrier 1 and the impeller of fluid clutch 6 is T_r/T_s times the torque on the shaft from sun gear 2 to the runner of that clutch, and the torques from these sources are in the same direction. However, the torque from ring gear 2 has two paths, one going to the impeller and one to planet carrier 1. If part of the torque were placed on the impeller, it would tend to turn the runner in the same direction as the runner tends to turn as the result of the torque from sun gear 2. There is no opposition to the torque from sun gear 2 except that which can be provided through the clutch by the impeller, so sun gear 2 drives the runner and no torque flows from ring gear 2 to the impeller. When Tq_{in} is the torque provided the "new" input shaft by the automobile, the torque on the shafts of the ring and sun gears of gear set 2 (Condition 6, Chart 3) are:

$$Tq_{\text{out } r} = (Tq_{\text{in}})\,(1/[1 + T_s/T_r]) = 0.62\,Tq_{\text{in}}, \text{ and}$$
$$Tq_{\text{out } s} = (Tq_{\text{in}})\,(1/[1 + T_r/T_s]) = 0.38\,Tq_{\text{in}}$$

Thus, 38 per cent of the input torque is transmitted through fluid clutch 6 from sun gear 2. This percentage corresponds with and has the same value as that in power-on operation pertaining to the hydraulic and mechanical path.

The slip in clutch 6 causes the speed of sun gear 2 to be greater than that of ring gear 2, so B is less than 1.0 and the SF of gear set 2 is reduced accordingly. The slip of fluid clutch 5 causes the speed of sun gear 1 to be greater than that of ring gear 1, so B in this case is less than 1.0 and the SF of gear set 1 is reduced accordingly. Then, if the slip of one or both of the fluid clutches is not small,

$$TF_{\text{trans}} = 1.0, \text{ and}$$
$$SF_{\text{trans}} = \text{less than } 1.0$$

the value depending on the amounts of the slips.

11. 1st-Gear Power-On Operation (In Low Position)

Clutches 1 and 2 and the band are engaged. Fluid clutch 5 is empty. Clutch 1 and the band Hold sun gear 1 and ring gear 2, but these would be Held by overrunning clutches 7 and 8 if clutch 1 and the band were not engaged. Therefore, the power flow is the same as when in 1st gear with the transmission control lever in the Drive position, and

$$TF_{trans} = 3.81, \text{ and}$$
$$SF_{trans} = 0.262 \text{ to } 0.0$$

12. 1st-Gear Power-Off Operation (In Low Position)

The clutch and band conditions are the same as in § 11.

Ring gear 2 is Held by the band and gear set 2 is in Condition 6 in Chart 1, so

$$TF_{gear \ set \ 2} = 1/(1 + T_r/T_s) = 1/2.63 = 0.38$$

The TF of clutch 6 is substantially 1.0, so all of the output torque of gear set 2 is placed on planet carrier 1. Gear set 1 is in Condition 3 in Chart 1 since clutch 1 Holds the sun gear, and

$$TF_{gear \ set \ 1} = 1/(1 + T_s/T_r) = 1/1.45 = 0.69$$

Then,

$$TF_{trans} = (0.38)(0.69) = 0.262, \text{ and}$$
$$SF_{trans} = 3.82 \text{ to } 0.0$$

The transmission shifts to Second gear when the automobile speed is above a few miles per hour if the throttle is closed.

13. Second-Gear Power-On Operation (In Low Position)

Clutch 2 and the band are engaged. Fluid clutch 5 is full of fluid.

The torque flow is the same as in 2nd gear when the transmission control lever is in the Drive position, so (after clutch 5 is full),

$$TF_{trans} = 2.63, \text{ and}$$
$$SF_{trans} = 0.38 \text{ to a lower value}$$

14. Second-Gear Power-Off Operation (In Low Position)

Clutch 2 and the band are engaged and fluid clutch 5 is full of fluid.

Gear set 2 is in Condition 6 in Chart 1. Gear set 1 is in Condition 6 in Chart 3, or substantially in one of Conditions 13–18 in Chart 1 when the slip of clutch 5 is small. Then the torque factor of the transmission is the reciprocal of that in § 13, and

$$TF_{trans} = 0.38, \text{ and}$$
$$SF_{trans} = 2.63 \text{ to a lower value}$$

15. 1st-Gear Power-On Operation (In S Position)

Clutches 1 and 2 are engaged and fluid clutch 5 is empty.

In 1st-gear power-on operation in the Low position, the band and

191

overrunning clutch 8 and clutch 2 Hold ring gear 2. In the S position, clutch 8 and clutch 2 Hold ring gear 2. The conditions of operation of the gear sets, then, are the same in the two positions, and

$$TF_{\text{trans}} = 3.81 \text{, and}$$
$$SF_{\text{trans}} = 0.262 \text{ to } 0.0$$

16. 1st-Gear Power-Off Operation (In S Position)

Clutches 1 and 2 are engaged and fluid clutch 5 is empty.

Torque from the "new" input shaft rotates ring gear 2 freely in the forward direction since overrunning clutch 8 overruns. The TF of the gear set is zero, and the TF and SF of the transmission are 0.0.

17. Third-Gear Power-On Operation (In S Position)

Clutches 1, 2, and 3 are engaged and fluid clutch 5 is empty.

The conditions are the same as in 3rd gear in the Drive position, except that clutch 1 is disengaged in 3rd gear. Clutch 1 Holds sun gear 1, which is Held by overrunning clutch 7 when in 3rd gear. Therefore, the TF and SF of the transmission are the same as in § 7, or

$$TF_{\text{trans}} = 1.45 \text{, and}$$
$$SF_{\text{trans}} = 0.69 \text{ to a lower value}$$

18. Third-Gear Power-Off Operation (In S Position)

Clutches 1, 2, and 3 are engaged and fluid clutch 5 is empty.

Clutch 3 and fluid clutch 6 substantially lock together ring gear 2 and sun gear 2 when the slip of clutch 6 is small, so the TF of gear set 2 is 1.0 and the SF is slightly less than 1.0. When the slip is not small, the torque on planet carrier 2 passes through gear set 2 and fluid clutch 6 to planet carrier 1 with a TF of 1.0, and the SF of these parts of the system is determined by the slip of clutch 6. Gear set 1 is in Condition 3 in Chart 1 since sun gear 1 is Held, and

$$TF_{\text{trans}} = 1/(1 + T_s/T_r) = 0.69 \text{, and}$$
$$SF_{\text{trans}} = 1.45 \text{ to a lower value}$$

19. Reverse-Gear Power-On Operation

Cone clutch 4 is engaged and clutch 5 is empty.

Gear set 1 is in Condition 4 in Chart 1 since the sun gear is Held by overrunning clutch 7, and

$$TF_{\text{gear set 1}} = 1.45$$

Gear set 2 is in Condition 4 in Chart 3, and

$$TF_1 = Tq_{\text{out pc}}/Tq_{\text{in s}} = 1 + T_r/T_s = 2.63 \text{, and}$$
$$TF_2 = Tq_{\text{out r}}/Tq_{\text{in s}} = -T_r/T_s = -1.63$$

Gear set 3 may be considered to be in Condition 5 in Chart 1, and

$$TF_{\text{gear set 3}} = 1 + T_r/T_s = 3.43$$

The torque on the output shaft from planet carrier 2 is

$$Tq_{\text{out}} = (1.45\, Tq_{\text{in}})\, (2.63) = 3.81\, Tq_{\text{in}}$$

where

$$Tq_{\text{in}} = Tq_{\text{in trans}}$$

The torque on the output shaft from ring gear 2 is

$$Tq_{\text{out}} = (1.45\, Tq_{\text{in}})\, (-1.63)\, (3.43) = -8.11\, Tq_{\text{in}}$$

The total torque on the output shaft is the sum of these two components, or

$$(3.81 - 8.11)\, Tq_{\text{in}} = -4.30\, Tq_{\text{in}}$$

Then,

$$TF_{\text{trans}} = -4.30, \text{ and}$$
$$SF_{\text{trans}} = -0.233 \text{ to } 0.0$$

20. Reverse-Gear Power-Off Operation

Cone clutch 4 is engaged and clutch 5 is empty.

Torque in the reverse direction with respect to the direction of rotation of the crankshaft is applied to planet carriers 2 and 3 by the "new" input shaft. From § 19, the TF of the combination of gear sets 2 and 3 in power-on operation is $-4.30/1.45$, or -2.97. In power-off operation, the TF of the combination of these two gear sets is

$$TF_{\text{gear sets 2 and 3}} = -1/2.97 = -0.3367$$

The output torque of this combination is applied to planet carrier 1 through fluid clutch 6. Since the TF above is negative and the "new" input shaft rotates in the reverse direction with respect to the crankshaft, planet carrier 1 and sun gear 1 are driven in the forward direction and overrunning clutch 7 overruns. Gear set 1 is in one of Conditions 7–12 in Chart 1 and the TF and SF of the transmission are 0.0.

21. Neutral-Gear Operation (In Neutral or Park Position)

All clutches and the band are disengaged and clutch 5 is empty.

Gear set 1 is in Condition 4 in Chart 1 in power-on operation, and its TF is 1.45. The torque from planet carrier 1 to ring gear 2 is zero since clutch 3 is disengaged. The torque to sun gear 2 is zero since ring gear 2 is Free. Therefore, the TF and SF of the transmission are 0.0.

In power-off operation in the Neutral position, the torque from the "new" input shaft may flow to clutch 4 which is disengaged, so the torque on planet carrier 3 is zero. The torque also may flow to planet carrier 2 and to ring gear 2, from which it may flow to disengaged clutches 2 and 3 and to sun gear 3, which can provide no opposition torque since clutch 4 is disengaged. Then, the *TF* and *SF* of the transmission are 0.0.

22. Control System Considerations

The pattern of clutch and band engagements in the forward-drive transmission control positions and gears, and the corresponding transmission torque factors, are shown below. The symbol, *E*, indicates engaged, or full of fluid in the case of clutch 5 which causes it to be "engaged." *H* or *OR* means that the overrunning clutch Holds or overruns, and *NI* means that it is "not involved." Overrunning clutch 7 is considered as not involved in power-off operation in 1st gear when in the *D* position, or in power-off operation in 2nd gear when in the *D* or *S* position, since gear set 2 transmits (substantially) no torque to gear set 1. It is not involved in any gear in which clutch 1 is engaged since clutch 1 Holds sun gear 1, although the overrunning clutch may assist clutch 1 in this Holding. Overrunning clutch 8 is not involved when the band is engaged, but it may assist the band in Holding ring gear 2. Clutch 4 is not included in the listing since it is used in Reverse gear only and no other clutch or band is used in Reverse gear. Fluid clutch 6 is not included since it is assumed to be "engaged" at all times. Certain of the gear shifts shown in Table 15 illustrate the problems involved in the design of the control system.

TABLE 15. CONTROL SYSTEM PATTERN

		Clutch				Band	OR Clutch 7		OR Clutch 8		Trans TF	
CP	Gear	1	2	3	F5	Band	P-On	P-Off	P-On	P-Off	P-On	P-Off
D	1st		E				H	NI	H	OR	3.81	0.000
	2nd		E		E		OR	NI	H	OR	2.63	0.000
	3rd		E	E			H	OR	OR	OR	1.45	0.000
	4th		E	E	E		OR	OR	OR	OR	1.00	1.000
S	1st	E	E				NI	NI	H	OR	3.81	0.000
	2nd		E		E		OR	NI	H	OR	2.63	0.000
	Third	E	E	E			NI	NI	OR	OR	1.45	0.690
L	1st	E	E			E	NI	NI	NI	NI	3.81	0.262
	Sec.		E		E	E	OR	OR	NI	NI	2.63	0.380

When in the *D* position, up-shifts from 1st to 2nd gear and from 3rd to 4th gear are accomplished by filling clutch 5. In power-on operation, the only control system problem is that of filling the clutch at a rate which prevents a surge in the speed of the automobile due to excess engine speed and at a rate which prevents an undue delay in the shifting process. Power-on up-shifts from 2nd to 3rd gear involve the removal of fluid from clutch 5 and the engagement of clutch 3. If the control system causes these clutches to transmit torque simultaneously and to significant degrees, the conditions are the same as or similar to those in 4th gear. Therefore, the transmission transmits torque, the engine speed tends to decrease, and the automobile speed tends to increase, the amounts depending on the extents of the engagements. Conversely, if a period of simultaneous and substantial disengagement occurs, the transmission tends to be in 1st gear and the engine speed tends to increase. If it increases, the problem of excess engine speed becomes greater than it would be if the simultaneous disengagement did not occur.

No problems exist in power-off shifts from 4th to 3rd and from 2nd to 1st gear when in the D position since the power-off TF of the transmission is zero throughout the shift or becomes zero during the shift. Likewise, no problems exist in power-on down-shifts between these gears, assuming that the rate of emptying clutch 5 permits the engine speed to increase to the required values during the shifts. Power-on shifts from 3rd to 2nd gear in which the above simultaneous engagements of clutches 3 and 5 occur tend to prevent the increase in engine speed that should occur during the shift. Conversely, prolonged simultaneous disengagement would permit the engine speed to increase by a factor which is greater than the desired factor of approximately 2.63/1.45. Simultaneous engagements of clutch 3 and fluid clutch 5 in power-off shifts from 3rd to 2nd would tend to place the transmission in 4th gear and increase the TF of the transmission from 0.0 to 1.0 and might result in undue deceleration of the automobile. Simultaneous disengagements would not adversely affect the shifts.

Power-off up-shifts from 1st to 2nd to 3rd gear present no problems if the above simultaneous engagements do not occur since the TF of the transmission remains zero. In the shift to 4th gear, the TF becomes 1.0, so the filling of clutch 5 should be at a rate which prevents a sudden decrease in the speed of the automobile because of a low engine speed.

The Low position frequently is used in power-off operation to cause the engine to act effectively as a brake. Therefore, power-off shifts from 1st to Second gear in this position should not permit the power-off TF of the transmission to become zero during the shift. Thus, the control system should not disengage clutch 1 until clutch 5 has enough fluid to prevent a sun gear speed which is substantially greater than the speed of ring gear 1. Likewise, in power-off down-shifts, clutch 1 should be engaged when clutch 5 still has enough fluid to prevent this speed to occur. This may result in a certain amount of "fighting" between the clutches if the completion of the engagement of clutch 1 is made too rapidly.

One of the greater engine speed adjustment problems is that which results from power-off shifts at high automobile speeds from the *D* position to the *L* position in order to provide greater engine "braking power." The speed factor of the transmission, disregarding slip in the fluid clutches, changes from 1.0 to 2.63. Clutch 3 and the band should not be engaged simultaneously to significant degrees since they would "fight" one another and tend to cause the sun and ring gears of gear set 2 to be locked to the transmission case which, if accomplished, would stop the rotation of the rear wheels of the automobile. Neither should the clutch and band be disengaged simultaneously for more than a very short period of time since the transmission would be in 2nd gear with a *TF* of zero and the automobile speed would increase. The completion of the engagement of the band should be at a rate which permits the engine speed to be increased by a factor of approximately 2.63 without causing an unduly fast deceleration of the automobile.

Appendix A: Definitions

Several of the following definitions include the terms *input shaft* and *output shaft*. In some instances, the connecting link between a gear element and its power source, load, or holding object is not a shaft. In these cases, the word *shaft* in the definition should be considered as being replaced by another which describes the connecting link.

1. Directions of Rotation

FORWARD: Unless indicated otherwise, the forward direction is the direction of rotation of the input shaft of a gear system or other system or device, which also is the direction of the torque applied to that shaft by the power source. When there are two or more input shafts, it is the direction of rotation of input shaft number 1 and the direction of the torque applied to that shaft. The input, output, etc., shafts are identified in the Condition of Operation statement (§ 5, below).

REVERSE: The direction of rotation which is the opposite of the forward direction.

2. Gears

CIRCUMFERENCE: The circumference, in feet, of the circle which is used in determining the number of gear teeth on the gear per foot of circumference of the gear. It includes the points which are approximately halfway between the bases and the tips of the teeth.

DIAMETER: Two times the radius of the gear.

RADIUS: The distance, in feet, from the center of the gear to the circle in the definition of circumference.

SIMPLE GEAR, OR SIMPLE PINION (Figs. 1 and 11): A gear, or pinion, consisting of a single gear with one set of gear teeth.

COMPOUND GEAR, OR COMPOUND PINION (Figs. 1, 19, 23): A gear or pinion consisting of one piece of material which has two or more sets of

197

gear teeth, or two or more simple gears connected together so that they may be placed on the same shaft, or two or more separate simple gears secured to the same shaft.

3. Loads

LOAD: The mechanism which is connected to an output shaft of a gear system or other system or device. (In power-off operation [§ 7, below], the normal input shaft is the output shaft, or "new" output shaft.) In some instances, load is used in place of load torque. For example, the term *heavy load* means large load torque.

LOAD TORQUE: The torque the load presents to the output shaft of a gear system or other system or device. At any instant it is equal in magnitude (but of the opposite sign) to the torque required to drive the load at the output-shaft speed that exists at that instant plus the torque required to provide the acceleration of the output shaft that exists at that instant. (Deceleration is negative acceleration.)

4. Planetary Gear Sets

PLANETARY GEAR SET (Figs. 11, 17, 19, 21): A grouping of gears, usually in a sun-and-planet fashion, in which planet gears may rotate around their axes and around another axis such as the axis of the sun gear or gears. Additional gears may be used in the gear set, and in some instances a sun gear is not used. The planet gears in a set of planet gears are identical and perform the same functions. Usually, there are three or four planet gears in a set, but one, two, or more than four may be used.

SIMPLE PLANETARY GEAR SET (Figs. 11, 19, 21): A gear set which has only three elements (sun gears, planet carriers, and ring gears) that may be connected to external devices by shafts or other connecting links.

COMPOUND PLANETARY GEAR SET (Figs. 17, 23): A gear set which has more than three elements of the type in a simple gear set.

5. Planetary Gear Systems

SIMPLE PLANETARY GEAR SYSTEM: A planetary gear system employing only one planetary gear set which is operated with only one input shaft and one output shaft. The gear set may be a simple or compound set.

COMPOUND PLANETARY GEAR SYSTEM: A planetary gear system employing only one planetary gear set which is operated with more than one input or output shaft, or more than one of each. The gear set may be a simple or compound set.

DUAL, TRIPLE (ETC.) PLANETARY GEAR SYSTEM (Figs. 24, 29, 32, 33): A planetary gear system employing two, three, or more simple and/or

198

compound planetary gear sets with the output shaft(s) of one connected to the input shaft(s) of another. The entire system may have only one input shaft and one output shaft, in which case its speed and torque characteristics are those of a simple gear system.

CONDITION OF OPERATION: A statement of the use or status of each shaft or element of a gear set, exclusive of the planet pinions. At least one shaft must be used as the gear set input shaft and at least one as the gear set output shaft in any Condition of Operation.

6. Speed and Torque Factors

SPEED FACTORS: In simple gear systems or other systems or devices which have only one input shaft and one output shaft, the speed factor is the ratio of the speed of rotation of the output shaft and the speed of the input shaft, taking into consideration the directions of rotation, or

$$SF = S_{out}/S_{in}$$

In gear systems which employ the gear set shown in Fig. 11 and have two input shafts and one output shaft, the speed factors are

$$SF_1 = S_{out}/S_{in\ 1}, \text{and}$$
$$SF_2 = S_{out}/S_{in\ 2}$$

for a specified value of A where $A = \pm S_{in\ 2}/S_{in\ 1}$, and where S_{out} is the output shaft speed and $S_{in\ 1}$ and $S_{in\ 2}$ are the speeds of input shafts 1 and 2.

In gear systems which employ the above gear set and have two output shafts and one input shaft, the speed factors are

$$SF_1 = S_{out\ 1}/S_{in}, \text{and}$$
$$SF_2 = S_{out\ 2}/S_{in}$$

for a specified value of

$$B = S_{out\ 2}/S_{out\ 1}$$

where S_{in} is the input shaft speed and $S_{out\ 1}$ and $S_{out\ 2}$ are the speeds of output shafts 1 and 2, taking into consideration the directions of rotation.

Input shaft 1, output shaft 2, etc., are identified in the Condition of Operation statement (§ 5, above).

TORQUE FACTORS: In simple gear systems or other systems or devices which have only one input shaft and one output shaft, the torque factor is the ratio of the torque the output shaft applies to the load and the torque applied to the input shaft by the power source, taking into consideration the directions of the torques, or

$$TF = Tq_{out}/Tq_{in}$$

In gear systems which employ the gear set shown in Fig. 11 and have two input shafts and one output shaft, the torque factors are

$$TF_1 = Tq_{out}/Tq_{in\ 1}, \text{and}$$
$$TF_2 = Tq_{out}/Tq_{in\ 2}$$

where Tq_{out} is the torque the output shaft applies to the load and $Tq_{in\ 1}$ and $Tq_{in\ 2}$ are the torques that are applied to input shafts 1 and 2 by the power sources or source, taking into consideration the directions of the torques.

In systems which employ the gear set above and have two output shafts and one input shaft, the torque factors are

$$TF_1 = Tq_{out\ 1}/Tq_{in}, \text{and}$$
$$TF_2 = Tq_{out\ 2}/Tq_{in}$$

where Tq_{in} is the torque applied to the input shaft by the power source and $Tq_{out\ 1}$ and $Tq_{out\ 2}$ are the torques output shafts 1 and 2 apply to their loads, taking into consideration the directions of the torques.

7. Power Conditions

POWER-ON OPERATION: Operation in which the driving torque flows from an automobile engine to the transmission and, except in the Neutral or Park position of the transmission control device, from the transmission to the driven wheels of the automobile.

POWER-OFF OPERATION: Operation in which the driving torque flows from the wheels of an automobile to the transmission, and from the transmission to the engine if the torque factor of the transmission and clutch, or automatic transmission, is not zero.

Appendix B: Symbols

Following are the most commonly used meanings of the symbols employed in this book. In some instances other meanings apply, as indicated by the discussion.

A	The ratio, $\pm S_{\text{in }2}/S_{\text{in }1}$, or an area in square feet.
B	The ratio, $S_{\text{out }2}/S_{\text{out }1}$
C	A constant value, or a constant value under stated conditions.
D, D_s, etc.	Diameter in feet of an object, such as the diameter of a sun gear.
F	Free, meaning a gear or shaft is free to rotate against (substantially) no opposing torque.
F_c, etc.	A force on the indicated object, in pounds.
H	Held, meaning a gear or shaft is held stationary.
I, I_1, etc.	Input shaft or gear of a gear system or other system or device.
Imp	Impeller of a fluid clutch or fluid torque converter.
K	Fluid velocity factor, or flow factor.
M	Mass of an object or body of fluid, which is its weight in pounds divided by 32.2.
m	Mass per cubic foot of an object or body of fluid.
O, O_1, etc.	Output shaft or gear of a gear system or other system or device.
R, R_s, etc.	Radius in feet of a gear or other object.
Run	Runner of a fluid clutch.
$S, S_{\text{in}},$ $S_{\text{out}},$ etc.	Speed of rotation of a shaft or gear or other object, the input shaft speed, output shaft speed, etc., in revolutions per minute unless otherwise indicated.
SF_1, SF_2, etc.	Speed factor of a gear system or other system or device.
St	Stator of a torque converter.
T_{ft}	Number of gear teeth on a gear per foot of circumference of the gear.

T_s, T_1, etc.	Number of gear teeth on a sun gear, gear number 1, etc.
TF, TF_1, etc.	Torque factor of a gear system or other system or device.
Tq_{in}, Tq_{out}, $Tq_{in\ 1}$, etc.	Torque applied to a shaft or gear, etc., or by one of these to its load or other object, in pound-feet.
$Turb$	Turbine of a fluid torque converter.
U	The ratio, $\pm Tq_{in\ 2}/Tq_{in\ 1}$
V_{r1}, V_{c4}, etc.	Linear velocity, or tangential velocity, in feet per second, of a rotating body or volume of fluid in the radial planes or circulatory planes of a fluid clutch or torque converter, where the radial-plane velocity is the velocity in planes that are perpendicular to the axis of rotation of the clutch or converter, and the circulatory-plane velocity is in planes which include that axis. The numerals 1, 4, etc., indicate the radius to which the velocity pertains.
W, W_{turb}, etc.	Angular velocity in radians per second of the indicated object.

Appendix C: Nonplanetary Gear Equivalents of Planetary Gear Sets in Figures 11 and 17

1. Equivalent Planet Carrier Gear Teeth

The speed factors of the simple planetary gear set shown in Fig. 11 in simple system Conditions 1-6 (Chart 1) are:

Condition	SF
1	$-T_s/T_r$
2	$-T_r/T_s$
3	$1 + T_s/T_r$ or $(T_r + T_s)/T_r$
4	$1/(1 + T_s/T_r)$ or $T_r/(T_r + T_s)$
5	$1/(1 + T_r/T_s)$ or $T_s/(T_r + T_s)$
6	$1 + T_r/T_s$ or $(T_r + T_s)/T_s$

The simple nonplanetary gear systems shown in Fig. 34A–C have these speed factors. Likewise, they have the same torque factors as the gear set shown in Fig. 11. Therefore, the planet carrier and its pinions appear to have the characteristics of a simple gear with $T_r + T_s$ gear teeth. However, since an idling gear must be used in Fig. 34B and 34C in order to provide the proper sign of the speed factors, it appears that the number of gear teeth on the simple gear in Conditions 3-6 should be $-(T_r + T_s)$, which is a negative number of teeth. Such a gear, of course, cannot be made.

2. Differential Action of Planetary Gear Sets

The operations of the gear sets shown in Figs. 11 and 17 in simple systems indicate that differential actions occur in the gear sets. For ex-

Equivalent of Gear Set
in Figure 11 in
Conditions

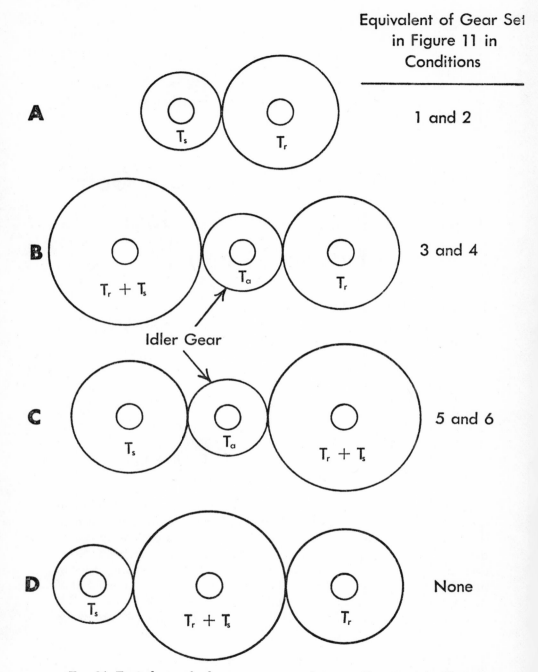

A	1 and 2
B	3 and 4
C	5 and 6
D	None

Idler Gear

FIG. 34. Equivalents of planetary gear set shown in Fig. 11. (1) All shafts rotate with their gears, but otherwise are stationary. (2) Shaft not used as input or output shaft is Free.

204

ample, power can be transferred from the sun gear shaft shown in Fig. 11, or from either of the sun gear shafts in Fig. 17, to the ring gear shaft even though the planet carrier shaft is held stationary. This cannot occur in the simple nonplanetary gear system shown in Fig. 34D. Differential actions also occur in compound systems which employ the gear sets shown in Figs. 11 and 17. Therefore, a nonplanetary gear system which is the equivalent of a planetary gear set must include a differential gear set.

A differential gear set is illustrated in Fig. 35. The following is a derivation of the relationships between the speeds of the three shafts of this gear set, where S_a, S_b, and S_c are the speeds of shafts a, b, and c, and S_p is the speed of the pinions.

(1) Assume that S_b is zero and S_c is not zero. Then,

    (a) $S_p = S_c$

    (b) Pinion rotations on their axes cause shaft a to have a speed component of $S_a = S_c$

    (c) Rotation of the differential case causes the pinions to revolve in the plane Q-Q at a speed of S_c, which in turn causes shaft a to have a speed component of S_c

    (d) The total speed of shaft a is the sum of the above components, or

$$S_a = S_c + S_c = 2\,S_c$$

(2) Assume that S_a is zero and S_c is not zero. Then, using reasoning such as that above,

$$S_b = 2\,S_c$$

(3) Assume that S_c is zero and that S_b is not zero. Then rotation of shaft b with a speed of S_b causes the pinions to rotate, which, in turn, causes shaft a to rotate with a speed of

$$S_a = -S_b$$

(4) Assume that S_c is zero and that S_a is not zero. Then, as in the above,

$$S_b = -S_a$$

(5) Assume that $S_b = S_a$ and that neither is zero or both are zero. Then the pinions do not rotate on their axes, and

$$S_c = S_b = S_a$$

(6) Assume that $S_b = S_c + S_x$, where S_x is any speed and is positive when it is in the forward direction. Then, the S_x component of S_b causes a $-S_x$ component in the speed of shaft a and (from (3) and (5), above),

$$S_a = S_c - S_x$$

205

From the above, the relationships of the speeds of the shafts of the differential gear set shown in Fig. 35, when $S_x = S_b - S_c$, are:

$$S_b = S_c + S_x,$$
$$S_a = S_c - S_x, \text{ and}$$
$$S_a + S_b = (S_c - S_x) + (S_c + S_x) = 2 S_c, \text{ or}$$
$$S_c = (S_a + S_b)/2$$

Certain special cases of the above are used frequently in the determination of the speed factors of systems employing the gear system shown in Fig. 36:

When $S_c = 0.0$, $S_b = -S_a$,
When $S_b = 0.0$, $S_a = 2 S_c$, and $S_c = S_a/2$,
When $S_a = 0.0$, $S_b = 2 S_c$, and $S_c = S_b/2$

3. Nonplanetary Gear Equivalents of Planetary Gear Sets

The gear system shown in the lower part of Fig. 36 is the equivalent of the gear set shown in Fig. 11 and the gear system in the entire illustration is the equivalent of the gear set shown in Fig. 17 insofar as the speed and torque factors are concerned. The internal gears connected to

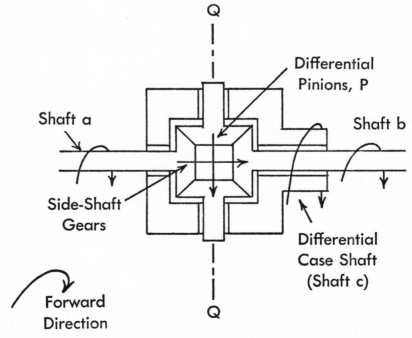

FIG. 35. Differential gear set. Numbers of gear teeth on side gears and pinions assumed to be the same, for sake of simplicity.

206

the side shafts of the lower differential have $(T_r + T_{s1})/2$ teeth, whereas the number indicated for those gears in § 1 is $(T_r + T_{s1})$. The difference is due to the fact that the speed factor of the differential, when one side shaft has zero speed, is two or one-half depending on the direction of the power flow through the differential. These same speed factors affect the number of teeth on the gears connected to the upper differential side shafts.

The use of the word, nonplanetary, in the titles of this chapter and this section may be considered to be erroneous since the differential gear sets shown in Fig. 36 are included in the definition of planetary gear sets in Appendix A. However, this form of gear set usually is called a *differential* rather than a *planetary* set.

The gear system shown in Fig. 36 cannot be constructed since $(T_r + T_{s1})/2$ and $(T_r + T_{s2})/2$ always are less than T_r. Therefore, gears with T_r teeth will not fit within the internal gears. The modifications of the system shown in Fig. 37 could be used if an equivalent system were built. However, Fig. 36 is used in the computations below in order to reduce the numbers of terms in the computations.

The following are examples of the determinations of the speed factors of the equivalent gear systems shown in Fig. 36 in which the conditions of the shafts are those stated in Charts 1 or 4. In all of the examples, the input shaft speed is assumed to be 1 rpm. Then, the speed of the output shaft in rpm is equal to the SF of the system since

$$SF = S_{\text{out}}/S_{\text{in}} = S_{\text{out}}/1.0$$

a. CONDITION 1, SIMPLE GEAR SET:

Sun gear 1 shaft — Input
Planet carrier shaft — Held
Ring gear shaft — Output

From § 2: $S_a = -S_b$ when $S_c = 0.0$
The speed of the ring gear shaft $= SF$, and is

$$(2\,T_{s1}/[T_r + T_{s1}])\,(-1)\,([T_r + T_{s1}]/[2\,T_r]) = -T_{s1}/T_r$$

which is the same as the SF of the planetary gear set shown in Fig. 11 in Condition 1 since sun gear 1 in Fig. 36 represents the sun gear in Fig. 11.

b. CONDITION 3, SIMPLE GEAR SET:

Sun gear 1 shaft — Held
Planet carrier shaft — Input
Ring gear shaft — Output

From § 2: $S_a = 2\,S_c$ when $S_b = 0.0$
The speed of the ring gear shaft $= SF$, and is

$$(2)\,([T_r + T_{s1}]/[2\,T_r]) = 1 + T_{s1}/T_r$$

207

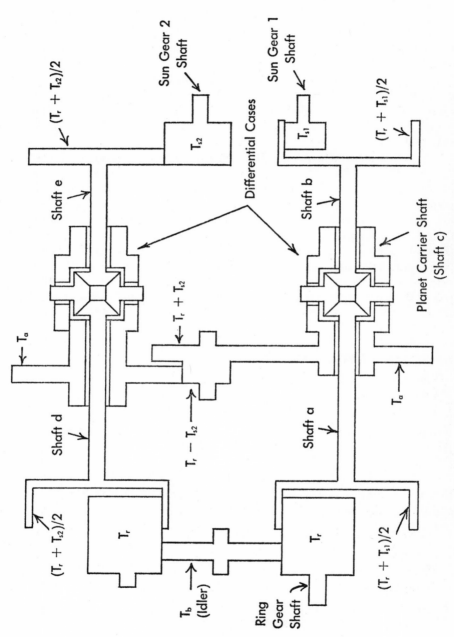

Fig. 36. Nonplanetary equivalents of Figs. 11 and 17. (1) T_a and T_b may be any convenient numbers. (2) Sun gear 1 is equivalent of sun gear in Fig. 11.

208

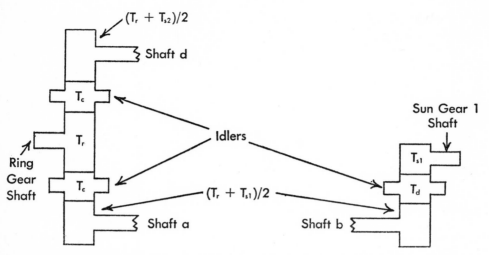

Fɪɢ. 37. Modifications of gear system shown in Fig. 36. (1) Remainder of gear system is same as shown in Fig. 36. (2) T_c and T_d may be any convenient numbers.

which is the same as the SF of the planetary gear set shown in Fig. 11 in Condition 3.

c. Cᴏɴᴅɪᴛɪᴏɴ 9, Cᴏᴍᴘᴏᴜɴᴅ Gᴇᴀʀ Sᴇᴛ:

Sun gear 1 shaft   — Held
Planet carrier shaft   — Input
Ring gear shaft   — Free
Sun gear 2 shaft   — Output

The speed of shaft d is

$$S_d = (2) \, ([T_r + T_{s1}]/[2 \, T_r]) \, (-T_r/T_b) \, (-T_b/T_r) \, ([2 \, T_r]/[T_r + T_{s2}])$$
$$= (2) \, (T_r + T_{s1})/(T_r + T_{s2})$$

The speed of the upper differential case is

$$S_{\text{upper case}} = (-T_a/[T_r + T_{s2}]) \, (+1) \, ([T_r - T_{s2}]/-T_a)$$
$$= (T_r - T_{s2})/(T_r + T_{s2})$$

From § 2: $S_e = 2 \, S_{\text{upper case}} - S_d$
Then,

$$S_e = (2) \, (T_r - T_{s2})/(T_r + T_{s2}) - (2) \, (T_r + T_{s1})/(T_r + T_{s2})$$
$$= (-2) \, (T_{s2} + T_{s1})/(T_r + T_{s2})$$

The speed of sun gear 2 = SF, and is

$$(-2) \, ([T_{s2} + T_{s1}]/[T_r + T_{s2}]) \, (-1) \, ([T_r + T_{s2}]/[2 \, T_{s2}]), \text{ or}$$
$$1 + T_{s1}/T_{s2}$$

which is the same as the SF of the planetary gear set shown in Fig. 17 in Condition 9.

WALTER B. LAREW, a retired Brigadier General in the United States Army, was born in 1904 in Elmdale, Indiana. In 1926, he received a B.S. in Electrical Engineering from Purdue University, later went on to study Communications Enginnering at Yale University.

While serving in the Army Signal Corps, Mr. Larew taught Military Science and Tactics at Cornell University. When World War II broke out, he served in the Canal Zone and India. As a Signal Corps officer he again saw action, this time in Korea in 1952 and 1953.

Between wars Mr. Larew directed the Communications School at the Air University in Alabama and was the author of material for both the Air Force and the Army. After his retirement from the Army in 1957, his continued interest in automotive engineering resulted in the writing of *Automatic Transmissions,* his first book on a nonmilitary subject.